Lubricant Blending and Quality Assurance

Lubricant Blending and Quality Assurance

By

R. David Whitby

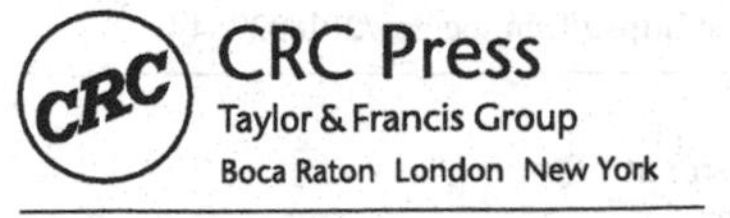

CRC Press is an imprint of the
Taylor & Francis Group, an **informa** business

CRC Press
Taylor & Francis Group
6000 Broken Sound Parkway NW, Suite 300
Boca Raton, FL 33487-2742

First issued in paperback 2021

CRC Press is an imprint of Taylor & Francis Group, an Informa business

ISBN-13: 978-0-367-78075-3 (pbk)
ISBN-13: 978-1-138-60593-0 (hbk)

Library of Congress Cataloging-in-Publication Data

Names: Whitby, R. David, author.
Title: Lubricant blending and quality assurance / R. David Whitby.
Description: Boca Raton, FL : CRC Press/Taylor & Francis Group, 2018.
Identifiers: LCCN 2018027644| ISBN 9781138605930 (hardback : acid-free paper) | ISBN 9780429466755 (ebook)
Subjects: LCSH: Lubricating oils--Additives. | Lubricating oil additives industry--Quality control. | Solution (Chemistry)
Classification: LCC TP691 .W44 2018 | DDC 665.5/385--dc23
LC record available at https://lccn.loc.gov/2018027644

Visit the Taylor & Francis Web site at
http://www.taylorandfrancis.com

and the CRC Press Web site at
http://www.crcpress.com

I dedicate this book to my wonderful wife, Jan, our talented daughters, Sarah and Helen, and our incredible grandchildren, Katherine, Andrew, Elliott and Natasha.

Contents

Preface

Many people, including those involved in the manufacturing, marketing and selling of lubricants, believe that blending lubricants is simply a matter of putting one or more base oils and several additives into a tank of some kind and stirring them around to mix them.

Blending lubricants that meet customer's demands requires much more than this. The correct ingredients of the right quality need to be used in precisely controlled quantities. The ingredients need to be tested prior to blending, and the finished products need to be tested following blending. The ingredients need to be stored and mixed under carefully controlled conditions. The finished lubricants need to be stored and packaged carefully and then delivered to customers correctly.

This book discusses all these issues, describes the different types of equipment used to blend lubricants, provides guidance on how best to use this equipment and offers tips and techniques to help avoid problems.

This book focuses on liquid lubricants. Greases are not be discussed, as their manufacture involves very different manufacturing procedures compared with those concerned with liquid lubricants.

Author

R. David Whitby is chief executive of Pathmaster Marketing Ltd., a business development consultancy for the international downstream oil, gas and energy industries, which he founded in 1992. Pathmaster Marketing has advised clients in the United Kingdom, France, Germany, Belgium, Denmark, Poland, the United States, Canada, Israel, Saudi Arabia, Iran, South Africa, Brazil, Singapore, Malaysia and Australia on business planning, business strategy, market development and technology commercialisation. Specialist sectors include lubricants, fuels, new energies and specialty chemicals.

An Australian by birth, David began his career with British Petroleum (BP), as a process chemist in a refinery in Western Australia. He worked for BP for 22 years in a number of management positions, including marketing and business development manager at Kalsep (an advanced separations company), business manager at BP Ventures, project leader for industrial lubricants at BP Research and marketing services officer at Duckhams Oils.

David was programme director for lubricants courses at the Oxford Princeton Programme, and he ran the Advanced Lubrication Training Programme for the UK Lubricants Association. He has written numerous papers and articles on lubricants and has chaired and lectured to international conferences and directed over 100 training lubricants courses in more than 30 countries.

In addition to running Pathmaster Marketing, David was non-executive chairman of Microbial Solutions Ltd., a start-up from the University of Oxford, from 2007 to 2015, and a non-executive director of the Sonic Development Company Ltd., from 1998 to 2003.

David has lived in Woking, Surrey, United Kingdom, for more than 35 years and is married with two daughters and four grandchildren.

1 Introduction

1.1 PURPOSE

Many people, including those involved in the manufacturing, marketing and selling of lubricants, believe that blending lubricants is simply a matter of putting one or more base oils and several additives into a tank of some kind and stirring them around to mix them.

Blending lubricants that meet customer's demands requires much more than this. The correct ingredients of the right quality need to be used in precisely controlled quantities. The ingredients need to be tested prior to blending, and the finished products need to be tested following blending. The ingredients need to be stored and mixed under carefully controlled conditions. The finished lubricants need to be stored and packaged carefully and then delivered to customers correctly.

This book discusses all these issues, describes the different types of equipment used to blend lubricants, provides guidance on how best to use this equipment and offers tips and techniques to help avoid problems.

This book focuses on liquid lubricants. Greases are not be discussed, as their manufacture involves very different manufacturing procedures compared with those concerned with liquid lubricants.

1.2 APPROACH

In order to manufacture high-quality lubricants, high-quality ingredients, suitable equipment, correct procedures and methodologies, standardised testing, appropriate packaging, correct storage conditions, a well-trained workforce and excellent management are required. All these subjects are discussed in this book.

Consequently, we approach this task in a methodical way. We start by considering the numerous types of base oils and additives that are used to manufacture lubricants. The base oils used in lubricants are classified into two broad groups: mineral oils and synthetic oils. Within each group there are significant differences between different base oils. There is no single "universal" base oil that will be suitable for use in all lubricant applications. This book illustrates and discusses all the main types of base oils used in lubricants.

Many hundreds of additives are used in different types of lubricants. This book explains why additives are used, the main types of additives and their functions and the very variable physical and chemical properties of different classes of additives. The fundamental importance of these differences to the operation of a lubricant blending plant is explained.

We then look into how and what goes into formulating a lubricant (selecting which base oils and additives to use) and why this can be important in making a product easier or more difficult to blend. The methodology of formulating, developing and testing a new or improved lubricant is complicated and time-consuming. Lubricant formulation chemists and engineers do not just select a base oil or additive because it happens to be available or low cost. Selecting which base oil(s) and additives to use for a specific lubrication application requires a great deal of experience, skill, testing and refinement.

Having discussed the ingredients and recipes for lubricants, we then look at how and why blending plants are designed. Many factors and a huge amount of market, technical, logistical and process information must be gathered and evaluated as inputs into the design of an efficient and profitable lubricant blending plant. This book describes and discusses these factors and the information required. We then look at the options available for lubricant blending plant layouts, with their advantages and disadvantages. Finally, the elements of the design process are considered, as well as the use of computer-aided design programmes for producing the detailed design plans and construction drawings.

Chapter 7 focuses on the types of equipment used in a lubricant blending plant. The advantages of each type of equipment are considered and the relationships between the various facilities are discussed. We look at the operation of each type of equipment, so that lubricant blending plant managers, supervisors and operators can gain insights into the efficient and effective use of the facilities in their plant. The chapter also discusses systems for automating the processes used to blend lubricants.

As with any commercial or industrial activity, difficulties can occur with blending lubricants from time to time. People frequently refer to "unforeseen circumstances". In the author's opinion, many of these circumstances are entirely foreseeable. One of the key tricks to managing and operating a successful activity is to avoid problems before they occur. Chapter 8, on avoiding problems, considers a number of foreseeable situations that can occur in a lubricant blending plant and how to guard against them happening. We look at lists of do's and don'ts to minimise risks before, during and after operations.

To start to consider product quality control, quality assurance and quality management, we look first at the testing and analysis of base oils and additives in blending plants. Manufacturing high-quality products requires high-quality raw materials, together with effective process management and control. This book describes the tests that a lubricant blending plant can use to evaluate the physical and chemical properties of the raw materials that will be used. It also provides guidance on the specifications that can be used for these base oils and additives, as well as the relationships that can be applied to the companies that supply these raw materials.

Following on, Chapter 9 looks at the testing and analysis of the lubricants that have been blended. Hundreds of different tests can be used to assess the properties and performances of lubricants. Many are used in the formulation and development of new or improved lubricants, and they are likely to be expensive and time-consuming. As a consequence, they are of little practical value in a lubricant blending plant.

The tests used in a blending plant need to be quick and comparatively low cost, to enable products to be packaged and delivered to customers as soon as practical. Many of the tests used to evaluate the chemical or physical properties of base oils and additives are also applicable for blended lubricants. However, there are a number of tests that are only applicable to finished lubricants, so these are described. Blending plant managers, supervisors and operators need to know and understand what happens in a blending plant laboratory.

In order to deliver lubricants that meet customers' requirements, the supply chain must be able to control the quality of the products it supplies. The first problem is to define what is meant by quality. The second problem is to establish methods for measuring it, and the third problem is to implement strategies and plans to control it. Chapter 11, on quality control, focuses on all three issues. It provides definitions and advice on methodologies for quality control and quality assurance. Quality management, a different subject, is the focus of Chapter 14. We look at controlling quality before, during and after blending. We also discuss external organisations that monitor the quality of lubricants in the market. At the end of the chapter, we present and discuss an established method to help the quality control and quality assurance processes, using a coding system for raw materials and lubricant formulations.

High-quality blended lubricants need to be packaged correctly into the various types of containers that will be used to deliver them to customers. Some containers are very large, while others are very small. This book describes the numerous types of packages used for lubricants, together with the advantages and disadvantages of some of them. Filling lubricants into these containers is also very important for the effective and efficient delivery of the products to users. Various methods used to fill lubricants into containers are presented and discussed. For many of the lubricants used by customers, the packaging forms an integral and important part of the marketing and branding of the products. This book discusses these aspects of lubricant packages. We also look at some aspects of the reuse or recycling of many of the packages for lubricants.

Storing lubricants correctly is just as important in a blending plant as it is in a customer's premises. The facilities and methods of operating a lubricant blending plant warehouse are presented and discussed. This includes warehouse management systems and automated storage and retrieval systems and their advantages and disadvantages. Large volumes of lubricants tend to be stored outdoors, while smaller volumes of lubricants and products that are sensitive to oxidation, moisture, dust, dirt, heat or cold tend to be stored indoors. Aspects of the storage of base oils and additives are be presented and discussed.

Producing and delivering high-quality products does not just happen. The entire process has to be managed. The strategies and activities required to manage product quality effectively and efficiently have been developed and improved over many years. The culmination of these developments and improvements was the publication of the ISO 9000 series of standards. This book looks at the principles and methodologies of Total Quality Management, with a specific focus on ISO 9000, 9001 and 9004.

This book is intended for everyone involved in supplying high-quality lubricants to customers. It is important that everyone in the whole supply chain understands the importance of the roles that each of the participating groups plays. Without the skills, experience, communication and understanding of all the people involved in manufacturing, marketing and delivering lubricants to customers, the chances of providing consistently high-quality products efficiently and profitably all the time are not high.

This book aims to enable all the participants in a company's supply chain for lubricants to achieve their goals.

2 Mineral Oil Base Oils: API Groups I, II and III *Properties and Characteristics*

2.1 INTRODUCTION

Mineral base oils are classified into two broad types, paraffinic and naphthenic, depending on the types of crude oils from which they are derived. Naphthenic crude oils are characterised by the absence of wax or have very low levels of wax. They are largely cycloparaffinic and aromatic in composition; therefore, naphthenic lubricant fractions are generally liquid at low temperatures without any dewaxing. Conversely, paraffinic crude oils contain wax, consisting largely of n-paraffins, which have high melting points. Paraffinic crude oils also contain iso-paraffins and some cyclo-paraffins and aromatics. (An explanation of these types of compounds is given in Section 2.4.) Waxy paraffinic distillates have melting or pour points that are too high for winter use, so the n-paraffins have to be removed by dewaxing. After dewaxing, the paraffinic base oils may still solidify, but at higher temperatures than do naph-thenic ones because their molecular structures have a more paraffinic "character". Paraffinic base oils are preferred for most lubricant applications and constitute about 90% of the world's supply.

2.2 BASE OIL NOMENCLATURE

Lubricant base oils have a number of acronyms that are used in the industry to describe them. American Petroleum Institute (API) Group I and Group II base oils (see Section 2.6 for an explanation of Groups I, II, III, IV and V) are first described by their viscosities, measured in Saybolt universal seconds (SUS) at 100°F. Typical nomenclatures are 70, 100, 150, 300, 500, 600 and 900, although different refineries manufacture and supply many other viscosity grades. The lower numbers refer to less viscous base oils, with the viscosities increasing as the numbers get bigger.

This nomenclature was developed in the United States at the beginning of the twentieth century, as one of the first attempts to categorise paraffinic base oils. The SUS unit was first proposed by George M. Saybolt, and the Saybolt universal viscometer was first standardised by the U.S. Department of Commerce in 1918.[1] The SUS viscosity of a liquid is the time taken for 60 ml of the liquid, held at a specific temperature (usually 100°F [37.78°C]), to flow through a calibrated tube. SUS viscosities can be converted to other measurements of kinematic viscosity (see Chapter 9).

These viscosity numbers are followed by either "SN", in the case of API Group I base oils, or "N", in the case of API Group II base oils. "SN" refers to "solvent neutral". Group I base oils are generally manufactured using solvent refining processes (see Section 2.3), while Group II base oils are generally manufactured using hydroprocessing. In both cases, the "N" refers to "neutral", in that the oils are neither acidic nor basic in character. Before the introduction of solvent refining processes, the vacuum distillate fractions of crude oil were treated with concentrated sulphuric acid to remove as many aromatic compounds as possible. The resulting acidic "base oils" had to be neutralised with a concentrated solution of caustic soda (sodium hydroxide). These base oils were not very oxidatively or thermally stable, hence the need to develop better methods of manufacturing base oils with improved properties.

The highest-viscosity base oils are called "brightstock". These oils are so named because they are clear and bright, a concept that is discussed in more depth in Chapter 9. Again, before the development of solvent refining processes, base oils had very dark colours and were often opaque. Solvent refining processes produce base oils that are light in colour and transparent.

Mineral oil base oils are sometimes referred to as medium-viscosity index (MVI), typically 60–80, or high viscosity index (HVI), typically 95–100. (See Chapter 9 for an explanation of viscosity index.) Very high viscosity index (VHVI) base oils have viscosity indices greater than 120. MVI base oils are almost always naphthenic in character, while HVI and VHVI base oils are usually paraffinic in character.

API Group III base oils are also categorised by their viscosities, but by their kinematic viscosities at 100°C, in centistokes (cSt). The most common viscosity grades of Group III base oils are 4, 6 and 8 cSt, although 5, 7 and 10 cSt viscosity grades are also available.

2.3 METHODS OF MANUFACTURING BASE OILS

Within a naphthenic or paraffinic type, base oils are distinguished by their viscosities and are produced to certain viscosity specifications. Since viscosity is approximately related to molecular weight, the first step in manufacturing is to separate out the lubricant precursor molecules that have the correct molecular weight range. This is done by distillation.

First, in a distillation unit operated at atmospheric pressure (atmospheric distillation) lower-boiling fuel products of such low viscosities and volatilities that they have no application in lubricants (naphtha, kerosene, jet and diesel fuels) are distilled off. The higher-molecular-weight components which do not vaporise at atmospheric pressure are then fractionated by distillation at reduced pressures (vacuum distillation) of from 10–50 mmHg. Thus, the "bottoms" from the atmospheric distillation column are fed to the vacuum distillation column, where intermediate product streams with generic names such as light vacuum gas oil (LVGO) and heavy vacuum gas oil (HVGO) are produced. These may be either narrow cuts of specific viscosities destined for a solvent refining step or broader cuts destined for hydrocracking to fuels and lubricant feedstocks.

The vacuum column bottoms may contain valuable high-viscosity lubricant precursors (boiling point greater than 510°C), and these are separated from asphaltic components (these are black, highly aromatic components that are difficult to refine) in a deasphalting unit. Deasphalting units separate asphalt from refinable components by solubility, and this is usually solubility in propane for the purposes of manufacturing base oils. This waxy lubricant feedstock is called deasphalted oil (DAO). Further refining of the DAO by dewaxing, solvent refining and hydrotreatment produces brightstock, which is a heavy (very viscous) base oil that is a "residue" (that is, it is not a distillate overhead). The DAO can also be part of the feed to a lubricant hydrocracker to produce heavier base oils.

The waxy distillates and DAO require three further processing steps to obtain acceptable base oils:

- Oxidation resistance and performance must be improved by removal of aromatic molecules (particularly polyaromatics), nitrogen and some of the sulphur-containing compounds.
- The viscosity–temperature relationship of the base oil (improve the viscosity index [VI]) has to be enhanced by aromatics removal to meet industry requirements for paraffinic oils.
- The temperature at which the base oil "freezes" due to crystallisation of wax must be lowered by wax removal, so that the oil in lubricated equipment can operate at winter temperatures.

There are two strategic processing routes by which these objectives can be accomplished:

- Processing steps that act by chemical separation. The undesirable chemical compounds are removed using solvent-based separation methods. Aromatic molecules are removed in a process called "solvent extraction". Wax molecules are removed in a process called "solvent dewaxing". Together, these methods are known as "solvent refining". The by-products (extracts and waxes) represent a yield loss in producing the base oil. The base oil properties are determined by molecules originally in the crude, since molecules in the final base oil are unchanged from those in the feed.
- Processing steps that act by chemical conversion. Components with chemical structures unsuitable for lubricants are wholly or partially converted to acceptable base oil components. These processes all involve catalysts acting in the presence of hydrogen; thus, they are known collectively as "catalytic hydroprocessing". Examples are the hydrogenation and ring opening of polyaromatics to polycyclic naphthenes with the same or fewer rings and the isomerisation of wax components to more highly branched isomers with lower freezing points. Furthermore, the chemical properties of existing "good" components may be simultaneously altered such that even better performance can be achieved.

Conversion processes are generally considered to offer lower operating costs, superior yields and higher base oil quality. In conversion processes, the eventual base oil properties reflect to some degree the molecules originally in the feed, but the extent of chemical alteration is such that products from different feedstocks can be very similar.

Separation processes are often depicted as "conventional" technologies, and these solvent refining processes currently account for about 46% of the world's paraffinic base oil production. Group I base oils are typically made by separation processes. Conversion processes, most usually called hydroprocesses, account for the remaining 56% and use catalytic hydroprocessing technology developed since World War II. This route has become particularly significant in North America, where more than 70% of base oil production uses hydroprocessing. Group II and Group III base oils are manufactured using hydroprocesses. Some companies have chosen to combine separation and conversion, since the latter has been developed in steps and opportunities for synergism and the reuse of existing hardware have been recognised. Some Group III base oils have been made using a combination of solvent refining and hydroprocessing.

A diagrammatic flowsheet for a typical solvent refining process is shown in Figure 2.1. In the conventional solvent refining sequence, a polar solvent selectively extracts aromatics, particularly those with several aromatic rings and polar functional groups. This gives an aromatic extract (the reject stream) and an upgraded waxy "raffinate" whose viscosity is less than that of the feed due to the removal of these polyaromatics. The major purposes of the extraction step are to reduce the temperature dependence of the viscosity (increase the VI) of the raffinate and improve the oxidation stability of the base oil. Since the raffinate still contains wax, which will cause it to "freeze" in winter, the next step (solvent dewaxing) removes the wax. Again, a solvent-based method is used, in this case crystallisation of wax. This reduces the temperature at which the oil becomes solid, essentially the pour point. If desired, the wax can subsequently be de-oiled to make hard wax for direct commercial sale. The base oil now has almost all the desirable properties; however, in a last step it is usually subjected to mild hydrogenation ("hydrofinishing") or (previously) clay treatment, which improves colour and performance by taking out a few percent largely composed of polyaromatics and nitrogen, sulphur, and any oxygen compounds.

In Figure 2.1, LN is the abbreviation for "light neutral" base oil, which has a relatively low viscosity; MN is "medium neutral", with an intermediate viscosity; and HN is "heavy neutral", with a relatively high viscosity. BS is the abbreviation for brightstock. The process flowsheet in Figure 2.1 is an illustration only. Many manufacturers of base oils make four different viscosity grades, as illustrated in Figure 2.1. Some manufacturers make only two or three different viscosity grades, while other manufacturers make five different viscosity grades.

It is worth noting here that processing steps for manufacturing Group I, II and III base oils are done in what is known in the industry as "batch mode". While the vacuum distillation column is operated continuously, to fill up the three storage tanks and provide feed to the propane deasphalting unit shown in Figure 2.1, the contents of only one of these tanks are then processed through the solvent extraction unit.

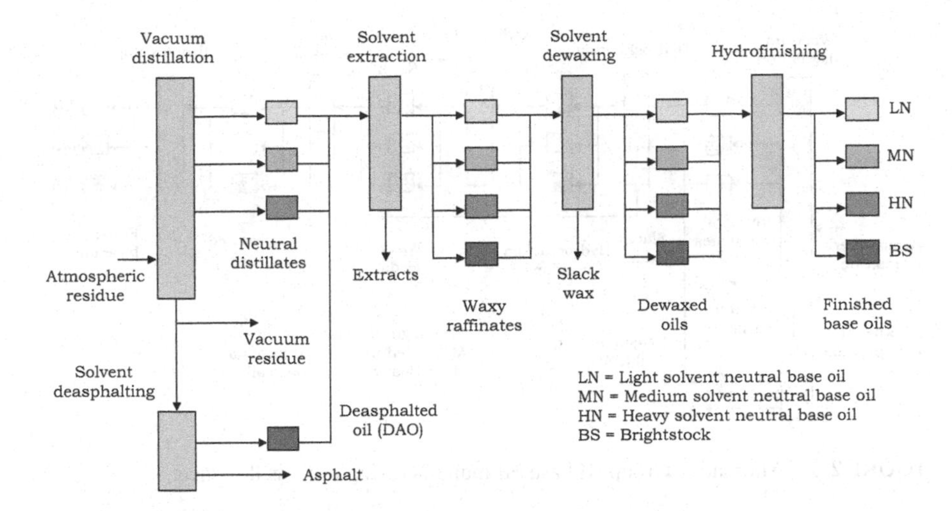

FIGURE 2.1 Group I base oil manufacturing process flowsheet.

At the same time, one of the storage tanks of waxy raffinates is processed through the solvent dewaxing unit and one of the storage tanks of dewaxed oils is processed through the hydrofinishing unit. Describing the precise processing methods used to operate these units is obviously outside the scope of this book, but more details can be found in books by Avilino Sequeira[2] or Thomas R. Lynch.[3]

In the hydroprocessing method, shown in Figure 2.2, catalytic hydrogenation in the first-stage lubricant hydrocracking unit saturates part of the feedstock aromatics by hydrogenating them to cycloparaffins. It also promotes significant molecular reorganisation by carbon–carbon bond breaking to improve the rheological (flow) properties of the base oil, again improving the VI. Usually in this stage, feed sulphur and nitrogen are both essentially eliminated. Some of the carbon–carbon bond breaking produces overheads in the form of low-sulphur gasoline and distillates. The fractionated waxy lubricant streams, usually those boiling above about 370°C, are

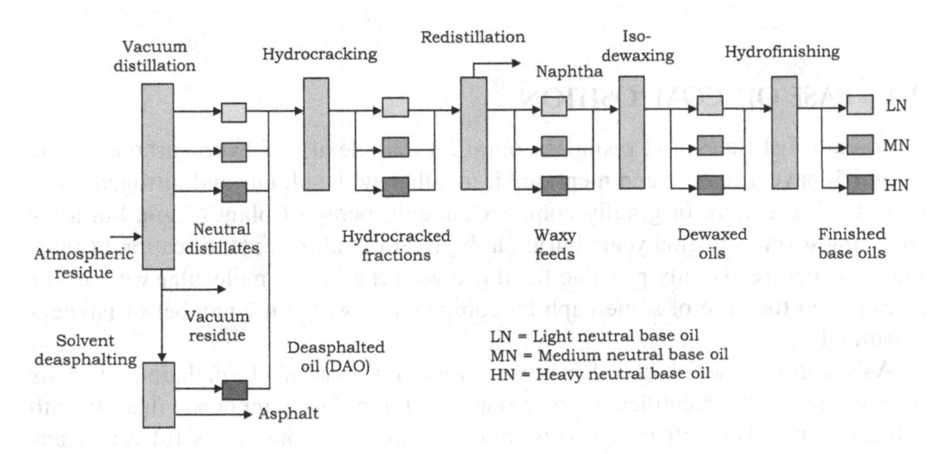

FIGURE 2.2 Group II or Group III base oil manufacturing process flowsheet.

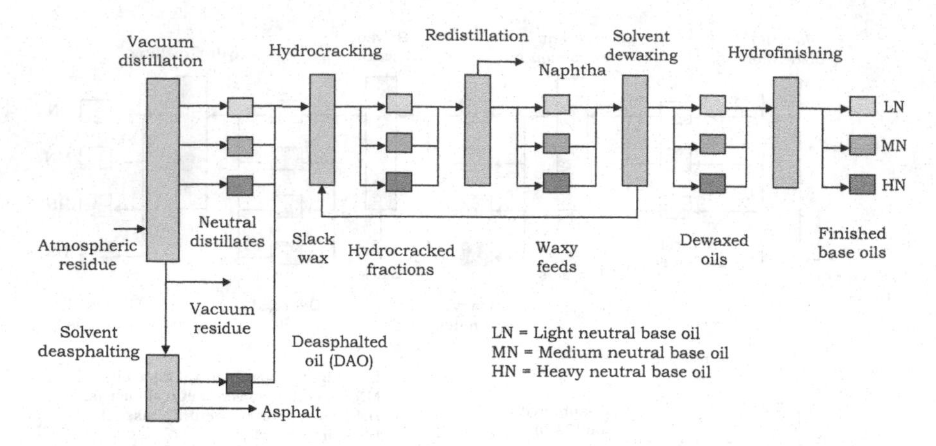

FIGURE 2.3 Alternative Group III base oil manufacturing process flowsheet.

then dewaxed, either by solvent dewaxing or, more usually, by catalytic hydroprocessing. In catalytic dewaxing, wax is cracked to lower-molecular-weight molecules. In isomerisation dewaxing ("iso-dewaxing"), wax is catalytically isomerised to low melting iso-paraffins in high yields. This also has a positive effect on VI. The final step in conversion processes is usually catalytic hydrogenation to saturate most of the remaining aromatics to make base oils stable for storage and to improve their performance. Base oils produced by this route are frequently water white, whereas solvent extracted oils retain some colour.

The process flowsheet for the combination of solvent refining and hydroprocessing sometimes used to manufacture Group III base oils is shown in Figure 2.3.

The advantages of the full hydroprocessing route are many: less dependence on supplies of expensive high-quality "lubricant" crudes, which the solvent refining process requires and which are increasingly in short supply; higher base oil yields; and lubricants that better (and in some cases exclusively) meet current automotive lubricant requirements.

2.4 BASE OIL COMPOSITION

Petroleum distillates and residues contain a complexity of hydrocarbons, some of which have already been mentioned, together with sulphur and nitrogen compounds. These were originally complex enough, being of plant origin, but after spending several million years buried at high temperatures, identification of individual structures is only possible for those with the lowest molecular weight, for example, in the case of some naphtha components, where the number of isomers is limited.

As a consequence, base oil manufacturers and lubricant formulation chemists have to rely on the identification of groups or "lumps" of compounds that fall into similar chemical classifications. This approach has been quite successful as a means of either separating or quantifying them by instrumental methods.

Beginning with the simplest, these chemical groups are:

- **n-Paraffins:** These are C_{18} and higher members of the n-paraffin homologous series, which are present in significant quantities in feeds and waxy intermediate streams with a boiling range of 315°C to 450°C, depending on the wax content of the feed. As the boiling point increases beyond 450°C, they become much less common. n-Paraffins are easily identified and quantified by gas chromatography because they give sharp peaks and can be concentrated in the slack wax fraction from solvent dewaxing. They are significant because they have high melting points and therefore increase the pour point of base oils. Base oils with low n-paraffin contents have low pour points.
- **Iso-paraffins:** These have n-paraffin backbones with alkyl side chain branches; on an iso-paraffin chain there may well be several branches of methyl groups or higher. Those iso-paraffins most similar in structure to n-paraffins (for example, single branches near a chain end) have higher pour points and will be removed by solvent dewaxing. Identification of individual members may be quite difficult. Iso-paraffins as a group are commonly said to have HVIs and low pour points, and confer good oxidation resistance. They are therefore a sought-after component in base oils. Polyalphaolefins (PAOs) are synthetic iso-paraffinic base oils (discussed in Chapter 3) that are of high commercial value because of their low pour points and excellent performance.
- **Cycloparaffins** (naphthenes): Cycloparaffins contain one or more cyclohexane (six carbon atoms) or cyclopentane (five carbon atoms) rings, or a combination thereof. If several rings are present, these are usually in the condensed form, presumably because of their natural origin. Mass spectroscopy of the saturates fraction can identify the number of rings and the percentage of the molecules having each number of rings. Alkyl substituents on the rings are branched and unbranched alkyl groups. Monocycloparaffins with 1,4 substituents are widely regarded as favourable structures, whereas polynaphthenes (3+ rings) are considered unfavourable for both VI and oxidation resistance.
- **Aromatics:** Basic structures have one to six or more benzene rings (six carbon atoms) with some of the carbon–hydrogen bonds replaced by carbon–carbon bonds of alkyl substituents. Generally, frequency declines with an increasing number of rings. Alkyl-substituted benzenes with 1,4-alkyl groups have HVIs and good oxidation resistance, whereas fused polyaromatic structures are undesirable.
- **Sulphur-containing compounds:** Because the source of the sulphur compounds is crude oil, they may be thiols, sulphides, thiophenes, benzo- and dibenzo-thiophenes and more complex structures. Solvent extraction reduces measured sulphur levels and therefore the content of sulphur compounds in solvent-refined base oils. Oxidation studies of solvent-refined base oils show that there appears to be an optimum level for sulphur compounds. Hydroprocessing will generally reduce the sulphur contents of base

oils to less than 10 ppm. The 4,6-dialkyl substituted dibenzothiophenes are the most resistant type of sulphur compounds to hydrotreating (due to steric hindrance), while thiols and sulphides are the most easily hydrogenated. Sulphur compounds constitute a poison to hydroisomerisation dewaxing catalysts and to nickel and noble metal catalysts and must be reduced to low levels in the feeds to those catalyst types.

- **Nitrogen-containing compounds:** Nitrogen compounds are mainly pyrroles, benzo- and dibenzo-carbazoles, pyridines and quinolines. Nitrogen compounds need to be minimised in finished base oils, since they contribute to colour instability. Hydrocracking vacuum gas oils reduces nitrogen levels to a few parts per million.
- **Oxygen-containing compounds:** Compounds containing chemically bound oxygen (such as furans and carboxylic acids) in lubricant feedstocks are seldom an issue in lubricant base oils and are generally disregarded. Hydroprocessing generally reduces their content in base oils to near zero.

2.5 MINERAL BASE OIL PROPERTIES AND CHARACTERISTICS

2.5.1 Overview

Base oils are manufactured to specifications that place limitations on their physical and chemical properties. These, in turn, establish parameters for refinery operations. Base oils from different refineries will generally not be identical, although they may have some properties, such as kinematic viscosity at a particular temperature, that are relatively similar. The test methods that are commonly used to determine the following properties and characteristics will be described in Chapter 9.

2.5.2 Appearance and Colour

With respect to "appearance", base oils should be "clear and bright" with no sediment or haze. Sediment can be caused by dirt or silt particles from storage, rust particles from processing or residual wax. Haze may be caused by silt or, more usually, by small droplets of free water. Clear and bright base oils tend to be indicative of correctly refined products.

Solvent-refined Group I base oils will retain some colour, as measured by the ASTM D1500 test method. Hydroprocessed Group II and Group III base oils, when hydrofinished at high pressures, are usually described as "water white", that is, colourless.

2.5.3 Density and Gravity

Knowledge of the density of a base oil is essential when handling quantities of it, particularly during blending finished lubricants. The values of density can also be seen to fit with the base oil types. Density increases with viscosity, boiling range, and aromatic and naphthenic content, and decreases as iso-paraffin levels increase and as VI increases. The density of a liquid also changes with temperature, so the temperature at which the density was measured must be stated. For petroleum products, including base oils, the standard temperature for measuring density is 15°C.

A base oil's relative density is usually quoted in units of grams per millilitre (g/ml) at the specified temperature. An alternative measure is the API gravity scale where, API gravity = 141.5/density – 131.5.

2.5.4 Viscosity and Viscosity Index

The viscosity of a base oil is measured in two ways: kinematic viscosity and dynamic viscosity. Viscosity is a measure of resistance to flow of a liquid, that is, its internal friction. Dynamic viscosity is the ratio between the applied shear and the shear rate, expressed in poise or centipoise (cP). Kinematic viscosity is the ratio of the dynamic viscosity to the density of the liquid, expressed in stokes or centistokes (cSt).

Kinematic viscosity is usually measured at 40°C and 100°C. Base oils are primarily manufactured and sold according to their kinematic viscosities at either 40°C or 100°C. Viscosity "grades" of blended lubricants are now defined by kinematic viscosity in centistokes (cSt) at 40°C. (As noted earlier, Group I and Group II base oils are still defined by the SUS scale at 100°F.) Higher-viscosity base oils are produced from heavier vacuum gas oils. For example, an oil with a kinematic viscosity of 100 cSt at 40°C is produced from an HVGO and cannot be made from an LVGO, since the molecular precursors are not present. As viscosity increases, so does the distillation midpoint.

Two methods are commonly used to measure the dynamic viscosity of a base oil:

- Brookfield low-temperature viscosity (ASTM D2983): This is the low-temperature shear rate apparent viscosity measurement between –5°C and –40°C and is reported in centipoise (cP).
- Cold cranking simulator: The apparent low-temperature viscosity of engine oils (ASTM D5293) correlates with the ease of low-temperature engine cranking, measured in centipoise rather than centistokes, and the temperature is always given, usually –25°C, –30°C or –35°C.

VI is a measure of the change of viscosity with temperature. It is well known that, for all liquids, viscosity decreases with increasing temperature. The higher the VI, the less the reduction in viscosity with increasing temperature. For many lubricants, higher VIs are generally preferred. VI is calculated from measurements of kinematic viscosity at 40°C and 100°C. The minimum VI for a paraffinic base oil is 80, but in practice the VI of a Group I base oil is 95–100 and for a Group II base oil it is 100–105. Naphthenic base oils may have VIs around zero, although they are more usually 60–80.

The conventional solvent refining route produces base oils with VIs between 95 and 100. Lower raffinate yields (higher extract yields) in solvent refining mean higher VIs, but it is difficult economically to go much above 105. In contrast, hydroprocessing enables a wide VI range of 95–140 to be achieved, with the final product VI depending on feedstock VI, first-stage reactor severity and the dewaxing process. Dewaxing by hydroisomerisation gives the same or higher VI relative to solvent dewaxing. To obtain a VI greater than 140, the feedstock generally must be either a petroleum wax or a Fischer–Tropsch (gas-to-liquid [GTL]) wax.

2.5.5 Pour Point and Cloud Point

The pour point of a base oil indicates the temperature at which it no longer flows under the influence of gravity. For paraffinic base oils, pour points are usually between –9°C and –18°C, and are determined by the operation of the dewaxing unit. For speciality purposes, pour points can be much lower. The pour points of naphthenic base oils, which can have very low wax contents, may be much lower (from –30°C to –50°C). For very viscous base oils such as brightstocks, pour points may actually reflect a viscosity limit.

The cloud point of a base oil is the temperature at which wax crystals first form as a cloud of microcrystals. It is therefore higher than the pour point, at which crystals are so numerous that flow is prevented. Many base oil inspection sheets no longer provide cloud points. Cloud points can be 3°C to 15°C above the corresponding pour points.

2.5.6 Distillation Range

At one time, the distillation range of a base oil would have been carried out using an actual physical distillation, using either ASTM D86, a method performed at atmospheric pressure and applicable to very light base oils, or vacuum distillation according to ASTM D1160 for heavier ones.

Now, distillation is usually performed by gas chromatography using ASTM D2887 and the method is commonly called either simulated distillation (SimDis) or gas chromatographic distillation (GCD). This method is capable of excellent accuracy, repeatability and fast turnaround times and is normally automated. It is applicable to samples with final boiling points of less than 540°C. For very heavy samples, ASTM WK2841 can analyse samples with boiling points in the range of 174°C to 700°C (C_{10}–C_{90}).

2.5.7 Flash Point

The flash point of a base oil indicates the temperature at which there is sufficient vapour above a liquid sample to ignite. It is a significant feature in product applications where it is used as a common safety specification. Flash points are a reflection of the boiling point of the material at the front end of the base oil's distillation curve. Flash points generally increase with viscosity grade. High flash points for a given viscosity are desirable. Good fractionation and increased base oil VIs favour higher flash points.

2.5.8 Volatility

The volatility of lubricants has emerged over the last 20 years as a significant factor in automotive engine oils, from both performance and environmental perspectives.

The volatility of a lubricant is heavily influenced by the volatility (or volatilities) of the base oil (or base oils) of which it is composed. The volatility of a base oil is determined by the "front end" of its distillation range. Low volatility (minimal losses at high temperatures) reduces emissions, is beneficial for emissions catalysts, reduces oil consumption and helps prevent engine oil viscosity changes. Volatility is obviously affected by viscosity grade, but for a constant viscosity is determined in part by narrower cut fractionation and in part by VI.

2.5.9 Aniline Point

Aniline point is a measure of the ability of the base oil to act as a solvent and is determined from the temperature at which equal volumes of aniline and the base oil are soluble. High aniline points (approximately 100°C or greater) imply a paraffinic base oil, while low aniline points (less than 100°C) imply a naphthenic or aromatic oil.

2.5.10 Viscosity–Gravity Constant

The viscosity–gravity constant of a base oil is an indicator of base oil composition and solvency that is calculated from the density and viscosity according to ASTM D2501. It usually has a value between 0.8 and 1.0. High values indicate higher solvency and therefore greater naphthenic or aromatic content. This is usually of interest for naphthenic oils.

2.5.11 Refractive Index and Refractivity Intercept

The refractive index is used to characterise base oils, with aromatic ones having higher values than paraffinic ones. The value increases with molecular weight. The refractivity intercept is calculated (ASTM D2159) from the density (d) and refractive index (n) (both at 20°C) using the sodium D line (ASTM D1218), where

$$\text{Refractivity intercept} = n - (d/2)$$

It is a means of characterising the composition of the sample. Values range from 1.030 to 1.047.

2.5.12 Elemental Contents

Sulphur is present in all lubricant plant feedstocks vacuum distilled from crude oil, and its content may be up to several percent. Solvent refining removes some but not all the sulphur compounds, so such oils with no further treatment can contain up to several mass percent of sulphur. Hydrofinishing of solvent-refined oils can reduce this level substantially. Base oils manufactured by hydroprocesses have sulphur levels in the low parts per million (ppm) range, as sulphur compounds are relatively easily removed with severe hydroprocessing.

Like sulphur, nitrogen is present in all lubricant feedstocks, generally in the 500–2000 ppm range. These levels are reduced by solvent extraction and nearly eliminated by hydrocracking.

Aromatic compounds are predominantly monoaromatics in both feedstocks and products, but substantial levels of di- and tri-aromatics can be present in feedstocks. Aromatics, particularly polyaromatics, make base oil oxidation stability much worse and can be virtually eliminated by hydroprocessing. Solvent extracted oils still contain substantial levels of aromatics.

Average carbon-type distributions, paraffinic (Cp), naphthenic (Cn) and aromatic (Ca) can be measured in base oils. Aromatic carbons are those in aromatic rings. To give a simple example, toluene has six aromatic carbons and one paraffinic carbon, and for this the percent Ca is 86 and the percent Cp is 14. Hydroprocessed base oils have Ca contents close to zero. These parameters can be determined via the n-d-M method (ASTM D3238) and from viscosity–gravity constant and refractivity intercepts via ASTM D2140. Percent Ca can also be determined by nuclear magnetic resonance (NMR) spectroscopy (ASTM D5292). The result will depend on the method used.

2.6 API BASE OIL CLASSIFICATIONS

A framework in which base oils are differentiated from one another for the purpose of base oil interchanges came with the development of base oil categories by the API in 1989 (Table 2.1). There are five categories. Paraffinic base oils belong to Groups I, II and III, while PAOs are in Group IV and all other base oils, including naphthenic base oils, esters, polyisobutenes and polyalkylene glycols, are in Group V.

Only three criteria are involved in determining in which group a base oil belongs: the percentage of saturates, the sulphur content and the VI. The percentage of saturates (saturated hydrocarbons) encompasses paraffinic and naphthenic compounds that do not contain carbon–carbon double bonds. Those compounds that contain carbon–carbon double bonds, notably aromatic compounds, comprise the remaining percentage in the base oil.

TABLE 2.1
API Base Oil Classification

	Group				
	I	**II**	**III**	**IV**	**V**
Saturates content	<90% and/or	>90% and	>90% and	Polyalphaolefins	All other base oils
Sulphur content	>0.03% and	<0.03% and	<0.03% and		
Viscosity index	>80 to <120	>80 to 120	>120		

Source: American Petroleum Institute, API Engine Oil Publications API 1509, 16th edition.

More recently, the base oil manufacturing and lubricants industries have adopted two additional base oil classifications, Group II+ and Group III+. These are not official API categories. These terms are employed in manufacturing and marketing and are frequently referred to. Group II+ base oils have low sulphur and aromatics contents and VIs generally in the range 115–118. Hence, they are not quite Group III quality, but are certainly better than Group II. Similarly, Group III+ base oils have VIs that tend to be 140 or higher, so they are clearly superior to Group III base oils with VIs in the range 123–130. Some manufacturers of base oils consider products with VIs greater than 130. Clearly, a Group III base oil with a VI of 133 is better than one with a VI of 123. However, in the author's opinion, Group III+ base oils should have a VI greater than 140.

The criteria for Group I base oils place no limitations on sulphur content or percent aromatics, and in practice these base oils are essentially all solvent refined. To reduce the aromatics content to less than 10% generally requires catalytic hydroprocessing. The VI range here is broad (80–120) and in practice most solvent-refined base stocks produced have VIs of 95–105. Generally, a VI of 95 is about the market minimum. HVIs are expensive to obtain by solvent extraction since yields decline rapidly as VI rises.

2.7 COMPARISON OF MINERAL OIL BASE OILS

2.7.1 Group I Base Oils

The typical properties of four different viscosities of Group I base oils are shown in Table 2.2. The data show that as the kinematic viscosity increases, so do the relative densities, the flash points and the sulphur contents. The base oil's colours are a light straw, except for the brightstock, which has a slightly darker (light brown) colour. The viscosity indices and pour points are relatively similar, while the volatilities, as measured in the NOACK test (see Chapter 9), decrease as the kinematic viscosities increase. All four base oils are clearly in Group I, as evidenced by their sulphur contents and chemical compositions. (Cp plus Cn comprise saturates compounds.)

Comparisons of 100 SN, 150 SN, 300 SN, 500 SN and 150 BS base oils from different manufacturers are shown in Tables 2.3 through 2.7. For each of the viscosity grades, the different base oils' physical properties are quite similar, with a few exceptions. For example, base oil A2 in Table 2.3 has a slightly lower kinematic viscosity, a slightly lower VI and a slightly higher sulphur content than the three other base oils in the table. Similarly, base oil B2 in Table 2.4 has a slightly darker colour than the other four base oils in the table.

The data shown in Tables 2.3 through 2.7 demonstrate that similar viscosity grades of Group I base oils manufactured by different refiners tend to have comparatively similar physical and chemical properties. This is despite crude oil feedstocks being different for different refiners. (It is well known in the oil industry that no two crude oils have exactly the same chemical compositions.) The refining processes used to make Group I base oils allow refiners to produce similar products from different crude oil feedstocks. These base oil manufacturing processes have been developed and improved for more than 60 years.

TABLE 2.2
Typical Properties of Group I Base Oils

	Base Oil			
Property	**150 SN**	**300 SN**	**500 SN**	**150 BS**
Colour	<1.5	<1.5	<1.5	<4.0
Density	0.879	0.885	0.888	0.906
KV at 100°C, cSt	5.5	8.0	10.9	31.7
KV at 40°C, cSt	34.1	61.6	97.8	493
VI	95	95	95	95
Pour point, °C	–18	–18	–15	–15
Flash point, °C	205	238	265	280
Sulphur content, % wt	0.30	0.40	0.60	0.80
NOACK, % wt loss at 250°C	19	6	3	—
Composition, % wt				
Cp	53	36	27	18
Cn	36	43	45	49
Ca	11	21	28	33

Source: Pathmaster Marketing Ltd., from numerous industry publications.

TABLE 2.3
Properties of 100 SN Group I Base Oils

	Base Oil			
Property	**A1**	**A2**	**A3**	**A4**
Colour	<0.5	<0.5	<0.5	<0.5
Density	0.862	0.859	0.860	0.863
KV at 100°C, cSt	4.10	3.80	4.10	4.29
KV at 40°C, cSt	20.4	18.4	20.6	22.3
VI	100	92	99	95
Pour point, °C	–15	–18	–15	–12
Flash point, °C	196	—	196	193
Sulphur content, % wt	0.20	0.46	0.20	0.14
NOACK, % wt loss at 250°C	29	31	29	31

Source: Pathmaster Marketing Ltd., from industry publications.

2.7.2 Group II Base Oils

The typical properties of five different viscosity grades of Group II base oils are shown in Table 2.8. As with the Group I base oils, as the kinematic viscosities increase, so do the relative densities and the flash points. Meanwhile, the volatilities

TABLE 2.4
Properties of 150 SN Group I Base Oils

	Base Oil				
Property	**B1**	**B2**	**B3**	**B4**	**B5**
Colour	<0.5	<1.5	<0.5	<1.0	<0.5
Density	0.871	0.879	0.882	0.883	0.867
KV at 100°C, cSt	5.40	5.50	5.10	5.10	5.20
KV at 40°C, cSt	33.1	34.1	30.0	30.0	31.0
VI	96	95	95	95	96
Pour point, °C	−15	−18	−15	−12	−16
Flash point, °C	207	205	212	216	—
Sulphur content, % wt	0.27	0.30	0.45	0.58	0.32
NOACK, % wt loss at 250°C	19	18	20	17	19

Source: Pathmaster Marketing Ltd., from industry publications.

TABLE 2.5
Properties of 300 SN Group I Base Oils

	Base Oil			
Property	**C1**	**C2**	**C3**	**C4**
Colour	<0.5	<1.5	<1.0	<1.5
Density	0.877	0.880	0.878	0.885
KV at 100°C, cSt	7.92	8.37	8.05	8.00
KV at 40°C, cSt	59.6	65.3	62.0	61.6
VI	98	97	95	95
Pour point, °C	−12	−12	−12	−18
Flash point, °C	244	227	235	238
Sulphur content, % wt	0.20	0.31	0.35	0.40
NOACK, % wt loss at 250°C	5	4	5	6

Source: Pathmaster Marketing Ltd., from industry publications.

of the base oils decrease. All five base oils have water-white colours and very low sulphur contents, due to the hydroprocessing route by which they are manufactured, and similar viscosity indices.

The main differences between Group I and Group II base oils are lower sulphur contents, slightly higher VIs and slightly lower NOACK volatilities for equivalent-viscosity grades. No base oil manufacturer currently produces a Group II base oil with the viscosity of a Group I brightstock. The highest Group II viscosity is 600 N; Group I base oils are manufactured with viscosities of 750 SN, 900 SN and even 1000 SN. It is technically possible to manufacture a Group II "brightstock", but the

TABLE 2.6
Properties of 500 SN Group I Base Oils

	Base Oil			
Property	**D1**	**D2**	**D3**	**D4**
Colour	<1.5	<1.5	<2.0	<1.5
Density	0.882	0.888	0.885	0.884
KV at 100°C, cSt	11.2	10.9	11.1	11.0
KV at 40°C, cSt	98.0	97.8	99.0	98.5
VI	98	95	96	95
Pour point, °C	–12	–15	–12	–12
Flash point, °C	260	265	258	260
Sulphur content, % wt	0.20	0.60	0.35	0.38
NOACK, % wt loss at 250°C	3	3	4	3

Source: Pathmaster Marketing Ltd., from industry publications.

TABLE 2.7
Properties of 150 BS Group I Base Oils

	Base Oil			
Property	**E1**	**E2**	**E3**	**E4**
Colour	<4.5	<4.0	<4.0	<2.5
Density	0.896	0.906	0.902	0.898
KV at 100°C, cSt	34.9	31.7	32.0	31.8
KV at 40°C, cSt	567	493	499	482
VI	95	95	95	97
Pour point, °C	–12	–15	–9	–12
Flash point, °C	285	280	280	300
Sulphur content, % wt	0.52	0.80	0.65	0.40
NOACK, % wt loss at 250°C	—	—	—	—

Source: Pathmaster Marketing Ltd., from industry publications.

production costs are very high. This is because the hydroprocessing routes make it increasingly difficult to produce higher-viscosity base oils either technically or cost-effectively.

2.7.3 Group III Base Oils

The typical properties of five different viscosity grades of Group III base oils are shown in Table 2.9. As with the Group I and Group II base oils, as the kinematic viscosity increases, so does the flash point, while the NOACK volatilities decrease.

TABLE 2.8
Typical Properties of Group II Base Oils

	Base Oil				
Property	**100 N**	**150 N**	**220 N**	**350 N**	**600 N**
Colour	<0.5	<0.5	<0.5	<0.5	<0.5
Density	0.853	0.864	0.867	0.871	0.874
KV at 100°C, cSt	4.10	4.89	7.22	7.94	12.4
KV at 40°C, cSt	20.4	27.03	41.2	57.8	113
VI	102	103	103	103	101
Pour point, °C	−14	−12	−13	−12	−15
Flash point, °C	213	215	230	243	270
Sulphur content, % wt	<0.01	<0.01	<0.01	<0.01	<0.01
NOACK, % wt loss at 250°C	26	14	11	5	2

Source: Pathmaster Marketing Ltd., from numerous industry publications.

TABLE 2.9
Typical Properties of Group III Base Oils

	Base Oil				
Property	**4 cSt**	**5 cSt**	**6 cSt**	**7 cSt**	**8 cSt**
Colour	<0.5	<0.5	<0.5	<0.5	<0.5
Density	0.830	0.835	0.836	0.839	0.843
KV at 100°C, cSt	4.20	5.10	6.00	6.90	8.00
KV at 40°C, cSt	18.6	25.3	32.4	39.4	50.5
VI	127	126	133	135	128
Pour point, °C	−18	−15	−15	−18	−15
Flash point, °C	220	240	234	240	260
Sulphur content, % wt	<0.01	<0.01	<0.01	<0.01	<0.01
NOACK, % wt loss at 250°C	15	9	8	4	3

Source: Pathmaster Marketing Ltd., from numerous industry publications.

The sulphur contents are very low, and the pour points and relative densities are comparatively similar.

Group III base oils have much higher viscosity indices than either Group I or Group II base oils. Group III base oils are also water white in colour. Equivalent-viscosity Group III base oils have much lower NOACK volatilities than Group I and Group II base oils. The highest-viscosity Group III base oil being marketed currently is 8 cSt; this grade is approximately equivalent to 230 SN Group I. No Group III base oils with viscosities approximately equivalent to 500 SN, 600 SN or higher are currently being manufactured. Also, with current manufacturing technologies, it is not technically possible to manufacture a Group III brightstock which is clear and bright.

2.7.4 Group I, II and III Base Oils

A comparison of 4 cSt Group I, II and III base oils manufactured by different refiners is shown in Table 2.10. Base oils A1 and A2 are clearly in Group I, base oils F3 and F4 are in Group II and base oils F5 and F6 are in group III. This is because the saturates contents, sulphur contents and viscosity indices of each of the base oils determine which API Group they belong to.

A comparison of 4 cSt Group II and Group II+ base oils is shown in Table 2.11. Base oils G1 and G2 belong in Group II+, because their viscosity indices are between 113 and 118, higher than those of base oils F3 and F4. In other respects, the four base oils' pour points, flash points, sulphur contents and chemical compositions are quite similar. The only other major difference between base oils F3 and F4 and base oils G1 and G2 is that the latter two have lower NOACK volatilities. This is almost certainly due to their better fractionation in the vacuum distillation column following hydrocracking.

Similarly, a comparison of four 4 cSt Group III and Group III+ base oils is shown in Table 2.12. The physical properties of the base oils are very similar, except for base oil H2, which has a higher VI than base oils F5 and F6. However, the chemical

TABLE 2.10
Comparison of 4 cSt Group I, II and III Base Oils

	Base Oil					
Property	**A1**	**A2**	**F3**	**F4**	**F5**	**F6**
Colour	<0.5	<0.5	<0.5	<0.5	<0.5	<0.5
Density	0.862	0.859	0.853	0.851	0.829	0.828
KV at 100°C, cSt	4.1	3.8	4.1	4.0	4.0	4.0
KV at 40°C, cSt	20.4	18.4	20.4	19.8	18.6	18.1
VI	100	92	102	96	128	125
Pour point, °C	–15	–18	–14	–12	–27	–25
Flash point, °C	196	—	213	—	—	220
Sulphur content, % wt	0.20	0.46	0.001	0.003	0.008	0.001
NOACK, % wt loss at 250°C	29	31	26	31	16	15
Composition, % wt						
Cp	53	49	67	68	59	77
Cn	36	33	33	31	33	21
Ca	11	18	<1	1	8	2

Source: Pathmaster Marketing Ltd., from industry publications.

TABLE 2.11
Comparison of 4 cSt Group II and II+ Base Oils

	Base Oil			
Property	**F3**	**F4**	**G1**	**G2**
Colour	<0.5	<0.5	<0.5	<0.5
Density	0.853	0.851	0.845	0.848
KV at 100°C, cSt	4.1	4.0	4.0	4.2
KV at 40°C, cSt	20.4	19.8	18.5	20.1
VI	102	96	116	113
Pour point, °C	–14	–12	–23	–18
Flash point, °C	213	—	220	206
Sulphur content, % wt	0.001	0.003	0.001	—
NOACK, % wt loss at 250°C	26	31	17	15
Composition, % wt				
Cp	67	68	70	—
Cn	33	31	29	—
Ca	<1	1	<1	<1

Source: Pathmaster Marketing Ltd., from industry publications.

TABLE 2.12
Comparison of 4 cSt Group III and III+ Base Oils

	Base Oil			
Property	**F5**	**F6**	**H1**	**H2**
Colour	<0.5	<0.5	<0.5	<0.5
Density	0.829	0.828	0.839	—
KV at 100°C, cSt	4.0	4.0	4.5	4.0
KV at 40°C, cSt	18.6	18.1	20.8	16.6
VI	128	125	130	144
Pour point, °C	–27	–25	–24	–18
Flash point, °C	—	220	215	225
Sulphur content, % wt	0.008	0.001	0.001	0.002
NOACK, % wt loss at 250°C	16	15	12	16
Composition, % wt				
Cp	59	77	98.8	98
Cn	33	21	1.0	2
Ca	8	2	0.2	0

Source: Pathmaster Marketing Ltd., from industry publications.

properties of the four base oils are very different, even though all belong to Group III or Group III+ as a result of their viscosity indices, sulphur contents and saturates contents. Base oil F5 has a relatively high Ca content compared with the other three base oils, even though it is less than 10%; that is, its saturates content is greater than 90%. Base oil F6 has a lower Cp content and a higher Cn content than base oils H1 and H2.

The data shown in Table 2.12 tend to indicate that not all Group III base oils are similar, particularly with regard to chemical composition. The differences are likely to influence the relative performances of the base oils, especially under oxidising or thermally degrading conditions. The importance of this will become clear in the next section, which discusses the issues that may occur when one base oil is substituted by another base oil of the same viscosity grade.

A comparison of four 8 cSt Group III and Group III+ base oils is shown in Table 2.13. In the author's opinion, base oils I1, I2 and I3 clearly belong to Group III, while base oil I4 belongs to Group III+. In this case, the physical and chemical properties of the four base oils appear to be relatively similar.

Group II base oils can be made using solvent refining processes, but this tends to be uneconomic unless the feedstock vacuum distillates are of very good quality. Group III base oils will, in the vast majority of cases, be severely hydrotreated or moderately hydrocracked, since the low sulphur and high saturates limits (low aromatics of less than 10%) are otherwise difficult to attain. The majority of group II stocks produced have VIs of 95–105.

Group III base oils are differentiated by their VHVIs, which defines them as being products from either fuel hydrocracking units (which operate at high severities

TABLE 2.13
Comparison of 8 cSt Group III and III+ Base Oils

	Base Oil			
Property	**I1**	**I2**	**I3**	**I4**
Colour	<0.5	<0.5	<0.5	<0.5
Density	0.843	0.834	0.847	0.832
KV at 100°C, cSt	8.0	7.7	8.0	7.9
KV at 40°C, cSt	50.5	47.4	54.2	44.8
VI	128	128	124	148
Pour point, °C	–15	–15	–12	–18
Flash point, °C	260	250	248	—
Sulphur content, % wt	<0.001	<0.001	<0.001	<0.001
NOACK, % wt loss at 250°C	3	4	6	6
Composition, % wt				
Cp	—	99.7	99.6	100
Cn	—	0.1	0.2	0
Ca	—	0.2	0.2	0

Source: Pathmaster Marketing Ltd., from numerous industry publications.

and low residue yields) or hydrocracking or isomerising wax or highly waxy feedstocks. Group III+ base oils are sourced either from paraffinic wax hydroisomerisation or from GTL plants that produce Fischer–Tropsch waxes.

2.8 BASE OIL INTERCHANGEABILITY

During the late 1980s, the API, prompted by the major lubricant suppliers and lubricant additive manufacturers, realised that the costs of switching between different suppliers of base oils was becoming prohibitive, particularly for passenger car motor oils (PCMOs). Until then, when a lubricant manufacturer wanted to change base oil suppliers, all the laboratory and engine tests required to meet an API PCMO specification had to be repeated using the new base oil(s). As some of the engine tests cost several tens of thousands of dollars per test, the costs of switching soon added up.

With the introduction of the API SG specification in 1989, the API decided to also introduce guidelines for those tests that had to be repeated when changing from one base oil supplier and/or type. The initial 1989 guidelines are reproduced in Table 2.14.

The API introduced the guidelines as follows. "Not all base oils have similar physical or chemical properties or provide equivalent engine oil performance in engine testing. During engine oil manufacture, marketers and blenders have legitimate needs for flexibility in base oil usage. The API Base Oil Interchangeability Guidelines (BOI) were developed to ensure that the performance of engine oil products is not adversely affected when different base oils are used interchangeably by engine oil blenders".

TABLE 2.14
API Base Oil Interchangeability Guidelines, 1989

Base Oil in Original API-Licenced PCMO[b]	Tests[a] to Be Repeated When Base Oil Is Changed				
	Group				
	I	II	III	IV	V
Group I	IIIE, VE	IIIE	<30%, none >30%, all	<30%, none >30%, all	All
Group II	IIIE, VE	IIIE, VE	<30%, none >30%, all	<30%, none >30%, all	All
Group III	All	All	All	<30%, VE >30%, all	All
Group IV	All	All	<30%, VE >30%, all	None	All
Group V	All	All	All	All	All

Source: American Petroleum Institute.

[a] API sequence engine tests, applicable in 1989.

[b] Passenger car motor oil.

The initial guidelines were relatively easy to follow. For example, if a lubricant blender or marketer wanted to substitute a Group I base oil from supplier A with another Group I base oil from supplier B, the only tests that would have needed to be repeated in 1989 were the Sequence IIIE and Sequence VE engine tests. Conversely, if a lubricant marketer wanted to modify a formulation by substituting less than 30% of a Group II base oil with a Group III base oil, no additional testing would have been needed. However, substituting one Group III base oil with a different Group III base oil would have necessitated that all the API SG specification tests be rerun, in 1989.

The guidelines have been updated regularly and now include guidelines applicable to diesel engine oils in addition to gasoline engine oils. The guidelines define the minimum acceptable level of testing for interchanging a base oil that every marketer must perform as a condition for obtaining an API licence. It is understood that, when comparing base oil properties, the precision of the applicable test methods is taken into consideration. Use of the guidelines does not absolve a marketer of the responsibility for the actual performance of the licenced product sold in the marketplace. The licencee must still ensure all the engine and bench test results. The guidelines are subject to modifications based on new data, new or revised test methods and/or new performance specifications. The API stipulates that "the current Guidelines must always be used".

The current guidelines were issued in October 2012 and were revised in March 2015.[4] They can be obtained from the API website, www.api.org. Engine testing is not required when a single interchange base oil that meets the definition of Group I, Group II, Group III, or Group IV is used at less than or equal to 10% wt of the blended PCMO formulation. In some cases, higher percentages of Group III or Group IV may be substituted without further engine testing, as specified in the guidelines.

Currently, there are a number of separate tables of base oil interchange requirements for PCMOs:

- Sequence IIIE, IIIF, IIIFHD, IIIG, IIIGA and IIIGB tests.
- Sequence IVA tests.
- Sequence VE/VG tests.
- Sequence VIA/VIB tests.
- Sequence VID tests.
- CRC L-38/Sequence VIII tests.

For diesel engine oils, there are even more tables of requirements:

- Caterpillar 1M-PC tests.
- Caterpillar 1K tests.
- Caterpillar 1N tests.
- Caterpillar 1R tests.
- Caterpillar 1P tests.
- Passing Detroit Diesel 6V92TA tests.
- Mack T-9 tests.
- Mack T-8/T-8E tests.

- Mack T-10 tests.
- Mack T-12 tests.
- Cummins M11/M11 EGR tests.
- Cummins ISM and ISB tests.
- Roller Follower Wear Test (RFWT).
- Engine Oil Aeration Test (EOAT).

Discussion of these tests is outside the scope of this book, as are the specific requirements of each of the tables.

The API Base Oil Interchangeability (BOI) Guidelines are referenced and required in the American Chemistry Council (ACC) Code of Practice[5] and have been adopted in Europe in the Association Technique de l'Industrie Europėene des Lubrifiants (ATIEL) Code of Practice.[6] The BOI interchange guidelines are a core element of the ATIEL Code of Practice that describe the best practice and recognised procedures for developing and formulating engine lubricants that meet European Automobile Manufacturer's Association (ACEA) requirements. The interchange guidelines help to reduce the number of engine tests needed to validate the use of alternative base oils and viscosity grades when formulating engine oils that are able to meet the latest performance specifications set out by the ACEA sequences. ATIEL notes that "interchange guidelines are key, not only for avoiding unnecessary and expensive engine testing programmes, but critically for reducing new product development time. This enables new lubricant formulations to be brought rapidly to market to meet consumers' and manufacturers' needs, while ensuring the absolute technical integrity and performance of the formulations".

For blending plant managers, supervisors and operators, it is very important to know and understand the BOI guidelines. It is not possible to simply switch from one base oil supplier to another for automotive engine oils that have been formulated to meet either API, ACEA or International Lubricant Standardization and Approval Committee (ILSAC) specifications, unless authorised to do so by the lubricant marketer whose engine oils are being blended. Fortunately, the main manufacturers of automotive engine oil additives (see Chapter 4) routinely run a number of engine tests using base oils from different suppliers, in order to be able to provide reassurance to lubricant marketers and blenders about the interchangeability of base oils.

2.9 SUMMARY

Mineral oil base oils are exceedingly complex mixtures of hydrocarbons. Those manufactured using solvent refining tend to contain a wider range of molecular types and weights than those manufactured by hydroprocessing.

Many properties are used to characterise and assess mineral oil base oils. Some of these are used to define base oil quality or performance, and some are used to help control base oil manufacturing processes. Some tests are used for both.

Those base oil properties and characteristics of most interest to blenders of finished lubricants will be defined and discussed in more depth in later chapters, when specifications for purchasing and quality control of raw materials used to blend lubricants are considered.

REFERENCES

1. Herschel, W. H., Standardisation of the Saybolt Universal Viscosimeter, Technologic Papers of the Bureau of Standards, No. 112, U.S. Department of Commerce, 27 June 1918.
2. Sequeira, A. Jr., *Lubricant Base Oil and Wax Processing*, Marcel Dekker, New York, 1994.
3. Lynch, T. R., *Process Chemistry of Lubricant Base Stocks*, CRC Press, Boca Raton, FL, 2008.
4. API 1509, Engine Oil Licensing and Certification System, 17th Edition, September 2012, Addendum 1, October 2015 (Errata March 2015), Annex E, API Base Oil Interchangeability Guidelines for Passenger Car Engine Oils and Diesel Engine Oils.
5. American Chemistry Council Petroleum Additives Product Approval Code of Practice, January 2018, www.americanchemistry.com.
6. ATEIL Code of Practice, Issue 20, November 2016, https://atiel.org.

3 Synthetic Base Oils: API Groups IV and V *Properties and Characteristics*

3.1 INTRODUCTION

Mineral oils are completely satisfactory for the large majority of applications that they are required to satisfy. In the author's opinion, this happy situation is likely to continue for many years.

However, when more demanding or severe operating conditions are encountered, mineral oils suffer some limits to their performance. The more important of these limitations are:

- At temperatures below about –30°C, most mineral oils, even with added pour point depressant additives, start to solidify as the wax molecules that they contain precipitate from solution.
- At temperatures above about 180°C, the lower-molecular-weight components of most mineral oils will start to evaporate, causing the oil to become more viscous and altering its overall properties.
- When operating temperatures are high, mineral oils react more easily with the oxygen in the air, leading to the formation of gums and residues that thicken the oil.
- The viscosity and temperature properties of many mineral oils mean that they are too thick at low temperatures and too thin at high temperatures. Suitable compromises are not always possible.
- Although mineral oils are generally satisfactory as lubricants, under extreme conditions their boundary-lubricating and extreme-pressure properties are not good enough.
- Some of the minor components in mineral oils can be mildly toxic to some life forms. Although mineral oils are ultimately biodegradable, many environmentalists regard the rate of degradation as too slow.

When mineral oils are no longer able to perform satisfactorily, unconventional and synthetic oils having superior properties are available as alternatives.

3.2 CONVENTIONAL DEFINITIONS OF MINERAL AND SYNTHETIC BASE OILS

Before discussing unconventional and synthetic oil lubricants, a clear definition of the essential difference between these two classes of lubricant is required.

For the purposes of this book, a mineral oil is a hydrocarbon lubricant derived primarily from the conventional refining (typically vacuum distillation, solvent extraction and solvent dewaxing) of crude oil, as outlined in Chapter 2. A synthetic oil is one that has been manufactured by chemical synthesis, which is a process in which bigger molecules are made from smaller starting units. Although some of the starting units may be derived from components of crude oil, it is important to remember that it is the use of process chemistry to "synthesise" bigger molecules by joining the starting molecules together.

By this definition, very high viscosity index (VHVI) and ultra-high viscosity index (UHVI) base oils were most usually classified in the past as mineral oils, not synthetic oils. They are still prepared from crude oil, even though the refining process has included severe catalytic hydrogenation and redistillation of vacuum distillate fractions or catalytic isomerisation of petroleum waxes.

Shell developed a wax isomerisation process which produces an extra-high viscosity index (XHVI, which is a Shell trademark) base oil with properties superior to those of VHVI or UHVI base oils. In recent years, Shell has also developed a process, as have Sasol and ExxonMobil, for converting natural gas into highly paraffinic wax, which can then be isomerised into UHVI base oils which have virtually identical properties to those produced from crude oil. For this base oil, the distinction between mineral and synthetic has become distinctly blurred.

In 1999, the National Advertising Division (NAD) of the U.S. Council of Better Business Bureaus adjudicated on a dispute between Castrol and Mobil on the use of the word *synthetic* as a description of certain lubricants. The disagreement was over an advertising claim within the United States. Mobil objected (despite allegedly having marketed hydroisomerised American Petroleum Institute (API) Group III base oils as "synthetic" in Europe and elsewhere) that Castrol's hydroprocessed Sintec® was not synthetic. The NAD did not agree, ruling that Castrol's evidence, although not demonstrating its product's superiority, constituted a reasonable basis for the claim that the Castrol product, as then formulated, was a synthetic motor oil. It is now generally accepted worldwide that Group III base oils are synthetic in that they enable the manufacture of lubricants that provide "synthetic performance".

A mineral oil is therefore a "natural" product, made by separation processes; a synthetic oil is a "man-made" product, made by building big molecules from small ones. Although some synthetic oils will contain molecules that are very similar in structure and properties to those found in mineral oils, the essential difference between them is that a synthetic oil is composed of a narrow range of very similar or nearly identical molecules, while a mineral oil is composed of a wide range of approximately similar molecules. The chemical and physical properties of a synthetic oil are much more uniform and predictable than those of a mineral oil. This key difference is important in determining the cost, performance and applications for both types of oils.

Paraffinic base oils account for almost 90% of the world's production. They are most usually refined to give base oils which have a viscosity index (VI) of between 90 and 105. High viscosity index (HVI) base oils comprise API Groups I and II, while VHVI base oils comprise API Group III. Naphthenic base oils are low in wax content and hence have a low pour point, valuable in refrigerator and transformer oil production. Their high aromatic content affords high additive solvency but is unacceptable because of the health risk. Even when solvent refined, naphthenic base oils have lower VIs (85–95) than paraffinic base oils, so they are often classed as medium-viscosity index oils.

There is little need to describe the conventional processes for producing HVI base oils in much detail, since these have been described in Chapter 2. It is, however, worth making a few important points. In order to produce HVI oils, the vacuum distillate or deasphalted oil must be treated by solvent extraction, which will dissolve low viscosity index (LVI) aromatic materials, thus improving the VI of the remaining oil, called the raffinate. The extraction is carried out in a countercurrent extraction tower in a broadly similar way to that used in propane deasphalting. The solvents used are furfural and N-methyl pyrrolidone, the latter having largely replaced the older solvent phenol, which is highly toxic. In all three cases, the principles of the extraction are the same, and the units differ as a result of differences in the physical properties dictating contact with oil, solvent recovery and water removal. Control of the VI can be made by altering the extraction conditions, such as the amount of solvent circulated, so several grades of oil of different VIs can be made from a single basic cut.

Following solvent extraction, raffinates have been treated to give the desired VI, but almost all will contain sufficient wax to make them solid at normal atmospheric temperatures, so it is necessary to remove this wax to ensure that the oils will be free flowing under service conditions. In the solvent dewaxing process, the oil is dissolved in a solvent and the solution is then chilled until the wax crystallises out and can be filtered off. It is necessary to use a solvent to facilitate the filtration of the wax, because otherwise the oil would be too viscous to filter easily at the low temperature required. The solvents employed can be a mixture of methyl ethyl ketone and toluene, or a mixture of chlorinated hydrocarbons, such as methylene chloride and dichlorethane.

3.3 TYPES OF SYNTHETIC OILS

For convenience, synthetic oils are usually classified into three primary groups:

- **Synthetic hydrocarbons:**
  Olefin oligomers: Polyalphaolefins (PAOs), polyinternalolefins (PIOs).
  Polybutenes: Polyisobutenes (PIBs).
  Alkylated aromatics: Alkyl and dialkyl benzenes.
- **Organic esters:**
  Dibasic acid esters: Adipates, azelates, sebacates (diesters).
  Polyol esters: Trimethylpropanol (TMP), neopentyl glycol (NPG), pentaerythritol (PE).

Tribasic acid esters: Trimellitates.
Polymer esters.

- **Other synthetic oils:**
  Polyglycol ethers: Polyalkylene glycols (PAGs).
  Phosphate esters: Isopropyl phenol, tricresyl.
  Polyphenyl ethers.
  Halogenated hydrocarbons.
  Silicate esters.
  Silicones: Phenyl, methyl, dimethyl.

Each of these types has specific advantages and disadvantages, which are the subject of a complete, separate book.[1]

The use of synthetic oils as lubricants is not new. A very brief history of the development and introduction of synthetic oils in lubricant applications is shown in Table 3.1. In this chapter, we focus only on the properties of those synthetic oils that are used most frequently in blending high-performance lubricants:

- Polyalphaolefins.
- Diesters and polyol esters.
- Polyisobutenes.
- Polyalkylene glycols.

Manufacturing lubricants that are based on some of the more specialised synthetic oils listed above is most usually done by the specialist chemical companies that manufacture these oils. When it comes to blending operations, quality control and quality assurance, the same principles apply to these chemical companies' operations as will be described and discussed in later chapters of this book.

TABLE 3.1
Development of Synthetic Oils as Lubricants

Year	Research or Introduction
1927	Work by Dr. Zorn, IG Farben, on hydrocarbons and synthetic esters
1943	Commercialisation of silicones by Dow Corning and General Electric
1945	First PAG-based lubricants from Union Carbide
1951	First specification for a diester-based synthetic oil for gas turbines: MIL-L-7808
1953	Development of phosphate esters
1963	U.S. Navy specification for neopolyol ester–based oils for gas turbines: MIL-L-23699
1969	SINT 2000, semi-synthetic engine oil launched in Europe by Agip
1970	U.S. Air Force specification for PAO-based hydraulic fluid: MIL-H-83282
1977	Mobil 1, fully synthetic engine oil, launched worldwide

Source: Pathmaster Marketing Ltd.

3.4 PERFORMANCE ADVANTAGES OF SYNTHETIC OILS

The important performance advantages that specific synthetic oils demonstrate allow them to be used for those applications in which a mineral oil might struggle or be completely unsuitable. Particular performance advantages and what they mean in practice are summarised in Table 3.2.

None of the synthetic oils that are currently available possess all these performance advantages, but many demonstrate several of them. The most important performance advantages of the more commonly used synthetic oils are listed in Table 3.3.

TABLE 3.2
What the Advantages of Synthetic Oils Permit

Advantage	Permits
Lower pour point	Lower operating temperatures
Higher oxidation stability	Higher operating temperatures
Higher VI	Wider operating range, lower additive usage
Superior solvency	Longer drain periods
Lower volatility	Lower oil consumption, lower viscosity grades
Better lubricity	Lower viscosity grades
Biodegradability	Better environmental acceptability
Low toxicity	Food-grade approved lubricants
Clean burning	Specialist two-stroke oils
Flame resistance	Specialist hydraulic applications

Source: Pathmaster Marketing Ltd.

TABLE 3.3
Main Performance Advantages of Synthetic Oil Types

Synthetic Oil	Key Advantages
Polyalphaolefins	Very low pour point, VHVI, low volatility
Diesters	Very low pour point, VHVI, low volatility, biodegradable
Polyol esters	High operating temperature, low volatility, biodegradable
Polyisobutenes	Burn without residue, non-toxic
Polyalkylene glycols	Lubricity at low viscosity, non-toxic, can be soluble in either water or oil
Phosphate esters	Fire resistant
Alkyl benzenes	High dielectric strength, very low pour point

Source: Pathmaster Marketing Ltd.

3.5 PROPERTIES AND CHARACTERISTICS OF THE MAIN SYNTHETIC BASE OILS

3.5.1 Polyalphaolefins

The physical and chemical properties of PAO fluids make them attractive for a variety of applications requiring a wider temperature operating range than can normally be achieved by petroleum-based mineral oil products.

The typical physical properties of the main viscosity grades of commercially available low-viscosity PAOs are shown in Table 3.4. The main grades are all manufactured using 1-decene (C_{10}) as the starting material, and the final properties are determined by control of the reaction parameters and (depending on the manufacturer) selective distillation of the light oligomers. Some manufacturers of PAOs are now using C_{12} alphaolefins, mixtures of C_{10} and C_{12} or mixtures of C_8, C_{10} and C_{12} as the starting raw materials. The manufacturing process is shown in Figure 3.1.

The data show that all commercial grades of low-viscosity PAOs have relatively high VIs of around 135. (No VI is shown for PAO 2 because the VI is undefined for fluids having a kinematic viscosity [KV] of less than 2.0 cSt at 100°C.) The viscosity of an HVI fluid changes less dramatically with changes in temperature compared with the viscosity changes of an LVI fluid. A practical consequence this property is that PAOs may not require the use of VI-improving additives (VI improvers [VIIs]) in many applications. The presence of a VII is often less desirable, because some tend to be unstable toward shear. Once the VII begins to be torn apart by high shear stresses, the fully formulated fluid goes "out of viscosity grade".

TABLE 3.4
Typical Properties of Low-Viscosity Polyalphaolefins

	PAO				
Property	**2**	**4**	**6**	**8**	**10**
Colour	<0.5	<0.5	<0.5	<0.5	<0.5
Density	0.797	0.818	0.827	0.832	0.836
KV at 100°C, cSt	1.80	3.90	5.90	7.80	9.60
KV at 40°C, cSt	5.54	16.8	31.0	45.8	62.9
KV at –40°C, cSt	310	2,460	7,890	18,160	32,650
VI	—	129	138	140	134
Pour point, °C	–63	–70	–68	–63	–53
Flash point, °C	>155	215	235	252	264
Sulphur content, % wt	0.0	0.0	0.0	0.0	0.0
NOACK, % wt loss at 250°C	99	12	7.0	3.0	2.0
Composition, % wt					
Cp	100	100	100	100	100
Cn	0	0	0	0	0
Ca	0	0	0	0	0

Source: Pathmaster Marketing Ltd., from PAO manufacturers' data.

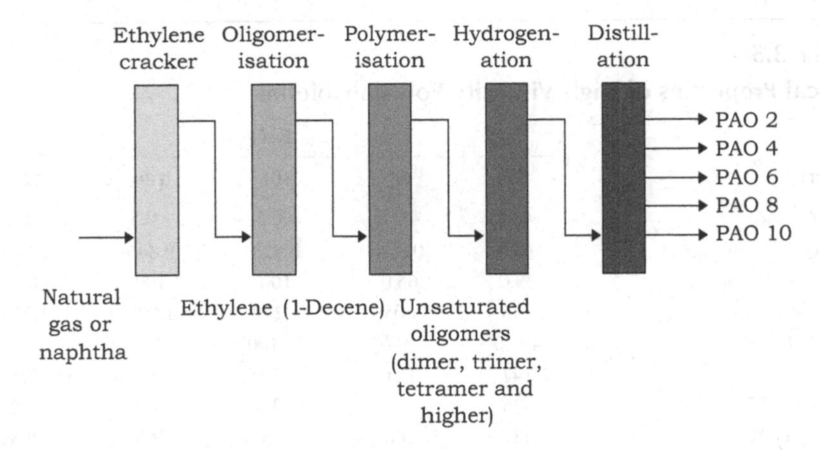

FIGURE 3.1 Polyalphaolefin manufacturing process flowsheet.

Several other important physical properties of low-viscosity PAOs are also shown in Table 3.4. All products have extremely low pour points and low viscosities at low temperatures. These properties make PAOs very attractive in applications in cold climates, which is where they were first used. At the other end of the spectrum, all but the 2.0 cSt product have low volatilities, as demonstrated by the low percentage loss of material at 250°C in the standard NOACK volatility test. Low volatility is important in high-temperature operations to reduce the need for "topping up" and to prevent a fluid from losing its lighter components and thus becoming too viscous at low or ambient temperatures.

The typical physical properties of conventional commercial high-viscosity PAO fluids are shown in Table 3.5. The four main high viscosity grades currently available are the 40, 65, 100 and 150 cSt fluids. As with the low-viscosity PAOs, these fluids have a very broad temperature operating range.

The excellent physical properties of the commercial PAO fluids are most readily apparent when they are compared directly with those of petroleum-based mineral oils. The fairest comparison is for fluids with nearly identical KVs at 100°C. The differences in both low- and high-temperature properties can then be compared. PAOs show markedly better properties at both high and low temperatures. At high temperatures, the PAO has lower volatility and a higher flash point. A relatively high flash point is, of course, often important for safety considerations. At the low end of the temperature scale, the differences are equally dramatic, with the highest degree of difference occurring in the low-temperature, low-shear regime, as is the case with dynamic viscosity as measured by either cold cranking simulator or Brookfield viscometer. The pour point of a 4 cSt PAO is −72°C, while that of a 100 SN Group I mineral oil is −12°C and that of a Group III mineral oil is −27°C.

The ability of PAOs to outperform mineral oil base oils of similar viscosity at both ends of the temperature spectrum can be easily understood if one compares their gas chromatographic spectra. A 4 cSt PAO is essentially a decene trimer with a small amount of tetramer present. The fine structure of the trimer peak is attributable to the presence of a variety of trimer isomers (same molecular weight, different

TABLE 3.5
Typical Properties of High-Viscosity Polyalphaolefins

	PAO				
Property	**40**	**65**	**100**	**100**	**150**
Colour	<0.5	<0.5	<0.5	<0.5	<0.5
Density	0.850	0.874	0.853	0.846	0.848
KV at 100°C, cSt	39.0	65.0	100	100	150
KV at 40°C, cSt	396	605	1,240	1,023	1,719
KV at 0°C, cSt	4,840	—	25,100	—	—
VI	147	181	170	192	205
Pour point, °C	–36	–45	–30	–42	–42
Flash point, °C	281	260	283	265	268
Sulphur content, % wt	0.0	0.0	0.0	0.0	0.0
NOACK, % wt loss at 250°C	0.8	0.7	0.6	0.6	0.5
Composition, % wt					
Cp	100	100	100	100	100
Cn	0	0	0	0	0
Ca	0	0	0	0	0

Source: Pathmaster Marketing Ltd., from PAO manufacturers' data.

structure). A 150 SN Group I base oil, on the other hand, has a broad spectrum of different molecular weight products. The oil contains low-molecular-weight materials that adversely affect the volatility and flash point characteristics. It also contains high-molecular-weight components that increase the low-temperature viscosity and linear paraffins that increase the pour point.

The chemical properties of PAOs are also superior to those of equivalent-viscosity mineral oils. The important properties of PAOs are:

- Superior thermal stability.
- Better response to oxidation inhibitors.
- Superior hydrolytic stability.

Many of the operations for which a functional fluid is required are carried out at elevated temperatures. For this reason, it is important that the fluid employed not be degraded under the operating conditions. The choice of an appropriate bench test, however, is often difficult. It is important that the test differentiate between thermal and oxidative degradation while simulating real-world operating conditions. It is also important that the test differentiate between thermal degradation and volatility. Some evaluations based on oven ageing or thermogravimetric analysis (TGA) have led to erroneous conclusions because the loss in sample weight and increase in viscosity could be attributed to volatilisation of the lighter components rather than chemical degradation.

In many of the oxidation and thermal stability tests used by lubricant formulators, PAOs and esters generally give better results than Group I or II mineral oils. It should be noted, however, that some oxidation and thermal stability tests show Group III base oils to be equal or sometimes superior to PAOs. The best performance is often achieved using a mixture of a PAO and a polyol ester. Diesters or polyol esters are commonly used in conjunction with PAOs in automotive engine oil formulations.

A high level of oxidative stability is essential to the performance of a functional fluid. In many applications, the fluid is required to perform at elevated temperatures and in contact with air. The results from attempts toward evaluation of fluids for oxidative stability, however, are often confusing. The results depend on the test methodology. Tests involving thin films tend to give different results than tests using bulk fluids. Not only the presence or absence of metals that catalyse oxidation is very important, but also the fact that different metals interact differently with different fluids. In addition, oxidative stability may be enhanced by the use of antioxidants, but different fluids respond differently to different antioxidants.

However, it must be noted that PAOs, like other base oils that are highly paraffinic in chemical composition, such as Group III base oils, have less inherent antioxidant capabilities than conventional Group I base oils.

For a functional fluid, the importance of inertness to reaction with water is important for a variety of reasons. Hydrolytic degradation of many substances leads to acidic products which, in turn, promote corrosion. Hydrolysis may also materially change the physical and chemical properties of a base fluid, making it unsuitable for the intended use. Systems in which the working fluid may occasionally contact water or high levels of moisture are particularly at risk. Also at risk are systems that operate at low temperatures or cycle between high and low temperatures.

The excellent hydrolytic stability of PAO fluids was reported as a result of tests conducted to find a replacement for 2-ethylbutyl silicate ester as an aircraft coolant or dielectric fluid used by the U.S. military in aircraft radar systems.

3.5.2 Diesters and Polyolesters

Synthetic ester lubricants, made from man-made raw materials having uniform molecular structures, have well-defined properties that can be tailored for specific applications. Ester lubricant base fluids are manufactured by the reaction of acids and alcohols, with the elimination of water. This reaction is reversible and the ester products can undergo hydrolysis, that is, reaction with water to regenerate the starting materials.

The types of acids and alcohols that can be used to manufacture esters for use in lubricant applications are shown in Table 3.6. The process for manufacturing acids from natural raw materials is shown in Figure 3.2. Reacting an acid with an alcohol will produce a single ester, while reacting mixed acids or mixed alcohols will produce mixed esters (Figure 3.3).

The ester linkage is an exceptionally stable one. Bond energy determinations predict that the ester linkage is more thermally stable than the carbon–carbon bond. The oxidative and thermal stability of ester base oils depends on the presence of hydrogen atoms on beta-carbon atoms and the number and type of hydrogen atoms present. As a result, linear acid esters are more stable than branched esters, while short-chain acid esters are more stable than long-chain acid esters.

TABLE 3.6
Acids and Alcohols That Can Be Used to Make Lubricant Base Oils

Acids	Chain Length	Alcohols	Chain Length
Monoacids		Linear alcohols	
Valeric	C_5	n-Hexanol	C_6
Capric	C_6	n-Heptanol	C_7
Heptanoic	C_7	iso-Heptanol	C_7
Capric	C_8	iso-Octanol	C_8
Pelargonic	C_9	2-Ethyl hexanol	C_8
Mixed cut	C_8+C_{10}	Mixed cut	C_8+C_{10}
Oleic	C_{16}	iso-Nonanol	C_9
		Mixed cut	C_9–C_{11}
Diacids		iso-Decanol	C_{10}
Adipic	C_6	Tridecanol	C_{13}
Azelaic	C_9		
Sebacic	C_{10}	Poly alcohols	
Dodecanoic	C_{12}	Neopentylglycol	NPG
		Trimethylolpropane	TMP
Dimer acids	C_{36}	Pentaerythritol	PE

Source: Pathmaster Marketing Ltd.

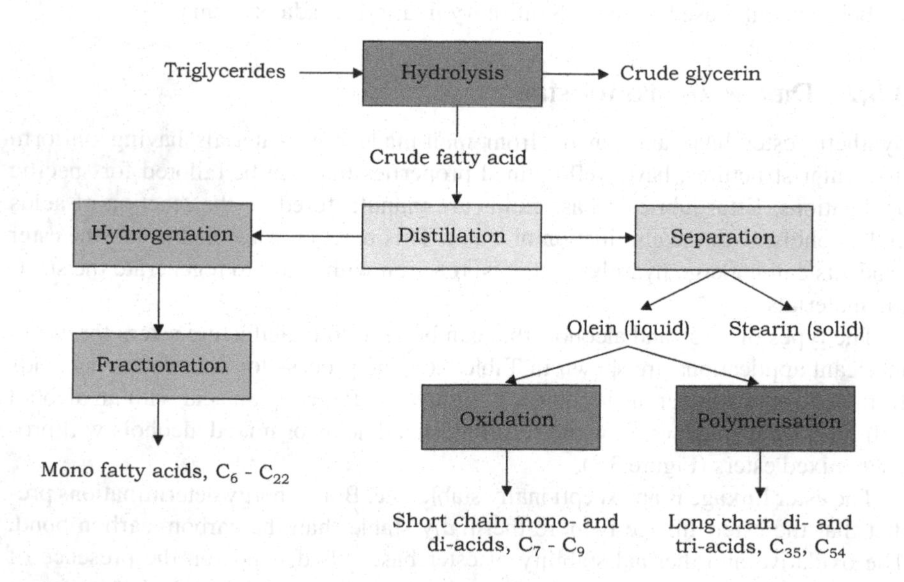

FIGURE 3.2 Manufacturing fatty acids.

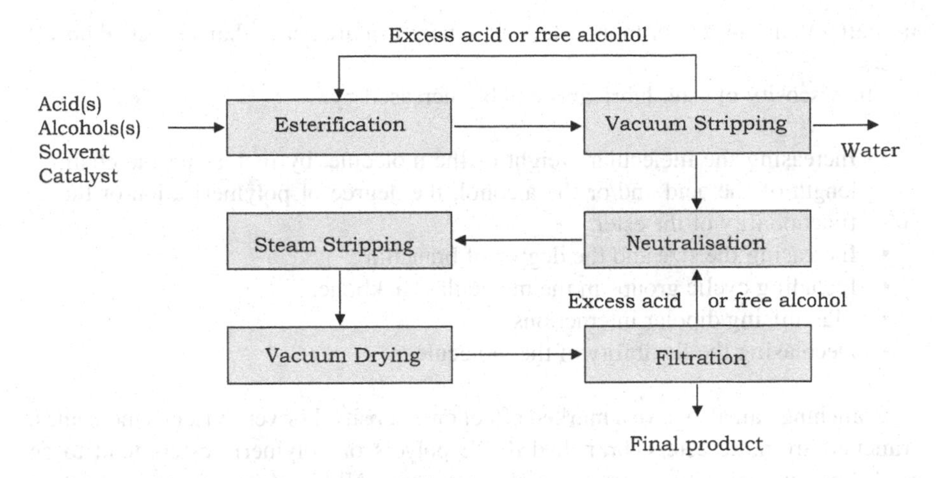

FIGURE 3.3 Manufacturing esters.

Esters made from linear acids generally have higher flash points than those made from branched chains or a mixture of linear and branched chains. Increasing molecular weight also increases the flash points. The volatility of esters depends on several parameters:

- Increasing molecular weight decreases volatility.
- Increasing the degree of branching increases volatility.
- Increasing polarity decreases volatility.
- Oxidative stability: Esters with low oxidative stability break down to form molecules of low molecular weight.

Although esters can undergo hydrolysis, the reaction between pure ester and water is very slow. Ester lubricants containing less than 500 ppm water can be stored for several years at ambient temperature and undergo essentially no reaction. For hydrolysis to occur at a significant rate, the following are required: some form of heat (>60°C), the presence of a catalyst (metal or acid) and a source of water (>100 ppm). In practice, the rate of hydrolysis is dependent on several factors, namely

- Temperature
- Presence of contaminants (particularly water or acidic species)
- Chemical structure of the lubricant (degree of branching)
- Specification of the ester (low acid and hydroxyl value, no residual catalysts)
- Presence, dose rate and type of additives

The biodegradabilities of esters can vary considerably, ranging from quite poor to excellent, depending on the specific test used to assess biodegradability. The main features that slow or reduce microbial breakdown are the position and degree of branching, the degree of saturation in the molecule, the presence of aromatic groups and high molecular weight. If high biodegradability is required, linear polyol esters are better than diesters or monoesters, which in turn are better than branched polyol esters.

The viscosity of ester lubricants can be increased by:

- Increasing the molecular weight of the molecule, by increasing the chain length of the acid and/or the alcohol, the degree of polymerisation or the functionality of the ester.
- Increasing the size and the degree of branching.
- Including cyclic groups in the molecular backbone.
- Maximising dipolar interactions.
- Decreasing the flexibility of the molecule.

Branching can also have a marked effect on viscosity. For very viscous molecules, branched aromatic esters, branched di-PE polyols or polymeric esters tend to be used. For low-viscosity esters, short-chain diesters, NPG polyols or monoesters are used.

The VI of an ester can be improved by increasing the acid or alcohol carbon chain length, increasing the linearity of the molecule and not using cyclic groups in the backbone and molecular configuration. The pour point of a diester or polyol ester can be improved by increasing the amount of branching, changing the positioning of the branch, decreasing the acid or alcohol carbon chain length or decreasing the internal symmetry of the molecule.

Esters made from mixtures of linear and branched chains have VIs between those of linear and branched, but have lower pour points than the esters obtained from either branched or linear chains. Pour point depressants can also be used, but they tend to be much less effective in esters than they are in mineral oil. Clearly, there is a trade-off between VI and pour point. For example, by increasing the linearity of the ester, the VI improves but the pour point deteriorates.

Esters have excellent compatibility with most other types of base oils. This means that changeover procedures can be relatively straightforward; esters can be used in machinery that previously used mineral oils, PAOs, PIBs and, in some cases, PAGs, subject to considerations with sealing materials (see below). Most additive technology is based on mineral oil, and it is therefore usually directly applicable to esters. Esters can be blended with mineral oil or natural oils (part synthetics) to boost their performance. Esters can be blended with other synthetics, such as PAOs, PAGs and PIBs, which gives esters great flexibility and unrivalled opportunities to balance the cost of different lubricant blends against performance.

Solubility problems can often result from the use of additives with PAOs and Group III base oils, due to their low polarity. This is especially true for VIIs. In many applications, esters are often combined with PAOs to overcome these solubility problems. As PAOs shrink seals and esters swell them, an optimum combination of the two can therefore be used to obtain a desired low-seal-swell target. The low friction of the ester component also compensates the poor fictional properties of the PAO. Ester/PAO combinations are therefore used in many applications.

Elastomers that are brought into contact with liquid lubricants will undergo an interaction with the liquid via diffusion through the polymer network. There are two possible kinds of interaction: chemical and physical. Chemical interactions of elastomers with esters are rare. During a physical interaction of an ester lubricant and an elastomer two different processes occur: absorption of the lubricant by the elastomer causing swelling and extraction of soluble components out of the elastomer causing shrinkage.

The degree of swelling of elastomeric material can depend on:

- Molecular size of the lubricant component. Generally the larger the lubricant, the smaller the swelling.
- Closeness of the solubility parameters of the lubricant and the elastomer. Generally, the "like dissolves like" rule is obeyed.
- Molecular dynamics of the lubricant. Linear molecules containing flexible linkages allowing rotation can diffuse into elastomers more easily than branched or cyclic ones.
- Polarity of the lubricant. It is known that several elastomers are sensitive to polar lubricants.

It is important to note that the processing of the elastomer can have a major impact on its performance. As esters can be efficient solvents, they have the potential to extract any substances used during the manufacture of the elastomer. Elastomers from different suppliers can be highly different in terms of the degree of cross-linking, fillers and process residuals in the elastomer. Therefore, any information on ester compatibility with elastomers in general should be confirmed by tests on the specific material.

Although many different combinations of acid(s) and alcohol(s) can be used to manufacture lubricant ester base oils, in practice only a small number of combinations are used commercially. These include:

- 2-Ethyl hexyl adipate.
- 2-Ethyl hexyl sebacate.
- iso-Decyl adipate.
- TMP sebacate.
- TMP oleate.
- NPG sebacate.

The main physical and chemical properties of some of these esters are shown in Tables 3.7 and 3.8. The data shown in Table 3.7 indicate that diesters have generally much lower pour points than mineral base oils and much higher VIs. As their viscosities increase, their NOACK volatilities decrease, as do their nitrile seal swell properties. These properties allow diesters to be used in lubricants, particularly automotive engine oils as volatility and seal shrinkage correction agents. (Highly paraffinic base oils, such as Group III base oil and PAOs, tend to shrink seals.) The biodegradabilities of diesters are excellent, except for the highest viscosity grades.

TABLE 3.7
Properties of Selected Diesters

	Diester					
Property	**A**	**B**	**C**	**D**	**E**	**F**
Density	0.91	0.91	0.91	0.91	0.91	0.91
KV at 100°C, cSt	3.6	4.3	4.8	5.6	6.7	12.6
KV at 40°C, cSt	13.9	17.4	20.2	23.8	36.7	85.7
VI	148	163	169	188	141	144
Pour point, °C	−70	−70	−60	−40	−52	−45
Flash point, °C	220	230	260	245	259	300
Nitrile seal swell, % vol.	17	10	8	5	3	0
NOACK, % wt loss at 250°C	14	9	6	6	4	1
Biodegradability, CEC L-103–12, %	100	100	100	90	80	40

Source: Pathmaster Marketing Ltd., from numerous manufacturers' data.

Similar observations apply to polyol esters, as indicated in Table 3.8. However, some polyol esters have higher pour points than others and some have lower VIs than diesters. Polyol esters can also be used as volatility and seal shrinkage correction agents. Their biodegradabilities are generally excellent.

3.5.3 Polyisobutenes

The structure of PIBs is essentially that of the iso-butylene repeat unit, with incorporation of low levels of other butene structures. The majority of polybutene polymer molecules contain a carbon–carbon double bond at the end of the polymer chains.

TABLE 3.8
Properties of Selected Polyol Esters

	Polyol Ester					
Property	**G**	**H**	**I**	**J**	**K**	**L**
Density	0.96	0.91	0.94	0.96	0.90	0.92
KV at 100°C, cSt	3.4	4.3	4.5	6.2	8.0	13.2
KV at 40°C, cSt	13.9	17.6	20.2	32.3	45.6	92.0
VI	120	160	140	144	148	143
Pour point, °C	−70	−22	−50	−10	−42	−30
Flash point, °C	235	245	250	285	280	300
Nitrile seal swell, % vol.	24	5	9	7	2	0
NOACK, % wt loss at 250°C	8	8	3	2	2	1
Biodegradability, CEC L-103–12, %	>90	>90	>90	>90	>90	>90

Source: Pathmaster Marketing Ltd., from numerous manufacturers' data.

A small minority of polymer chains are terminated with a carbon–halogen bond. The position of the double-bond end group is important in determining the ease with which it may undergo chemical reaction.

From carbon nuclear magnetic resonance (C-NMR) studies it has been shown that the positioning of the double bond in most polybutene grades gives rise to the cis- and trans-trisubstituted end group structure. An internal double bond may be present in some polymer chains, although these prove difficult to characterise. Polybutenes contain only one double bond per polymer molecule and therefore cannot undergo conventional cross-linking reactions. The composition of a given polybutene in terms of backbone structure and the positioning of the double bond at the end of the polymer chains can be influenced by the manufacturing conditions but is perhaps mostly dependent on feedstock composition and the overall manufacturing process (Figure 3.4).

Polybutenes having the predominantly disubstituted vinylidene structure at the end of the polymer chains are commercially available. The preponderance of the vinylidene end group structure gives improved chemical reactivity to the polymer in, for example, the maleinisation reaction with maleic anhydride to produce the important polybutenyl succinic anhydride derivative.

Polybutenes are commercially available as grades defined by a viscosity range. The nature of the polymerisation process results in each grade being made up of a distribution of polymer chains of different chain lengths or molecular weights. Polybutenes produced by cationic polymerisation have relatively narrow molecular weight distributions.

The main hydrogen types in polybutenes are secondary and primary. Incorporation of other butene structural types in the polymer chain introduces a small proportion of the more reactive tertiary hydrogen type. However, the main focus for any reactivity in the polybutene structure is the olefinic bond at the end of the polymer chain. Reaction of this olefinic group is normally achieved only under certain contrived conditions. Under normal storage conditions and for industrial applications, polybutenes can be considered to be stable products. The chemical stability of the polybutene polymers is well demonstrated by their retention of viscosity and tackiness, and

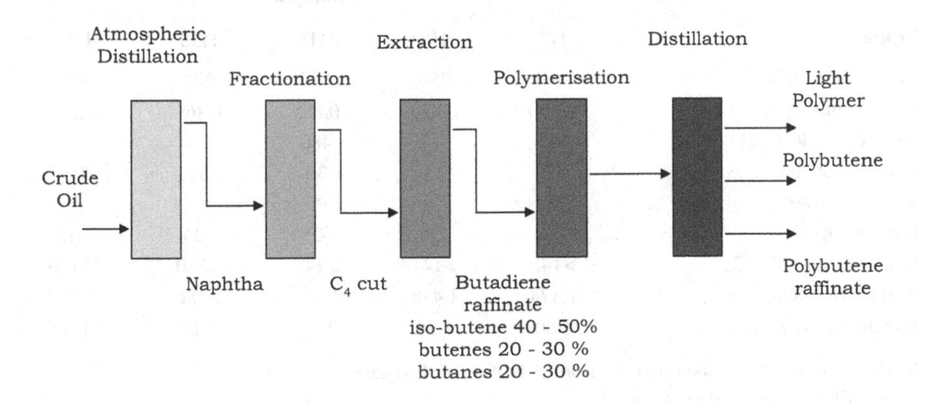

FIGURE 3.4 Polybutene manufacturing process flowsheet.

by their failure to harden, to become waxy, or to show any deterioration in colour upon storage for many years at ambient temperatures.

As produced, polybutenes are extremely pure materials containing extremely low levels of additional species, such as water, chlorine, and metals such as iron. The polymers are also free from nitrogen and sulphur species at detectable levels. Sensitive tests recognised for the detection of polycyclic aromatic hydrocarbons have failed to detect the presence of aromatic compounds in polybutenes.

Polybutenes are constructed of carbon and hydrogen and are non-polar in character. As such, they are normally soluble in non-polar solvents and insoluble in polar solvents. Polybutenes are insoluble in simple alcohols, esters and ketones. Solubility does increase, however, with higher homologues and as the molecular weight of the polybutene decreases. Polybutenes are fully compatible at all concentrations with low-, medium- and high-viscosity mineral oils of varying aromatic, paraffinic and naphthenic contents. Full compatibility has also been demonstrated with PAOs and alkyl benzenes. In most, but not all, cases polybutenes have been found to be compatible with synthetic ester oils. Incompatibility is found with silicone oils, except for some low-viscosity polybutene grades, and with PAGs.

Polychloroprene and nitrile rubbers show good resistance to polybutenes, as do fluoroelastomers such as Teflon and Viton. These materials are most suitable for the pump packing and seals of equipment used to handle polybutenes.

The KV of polybutenes is normally measured using a suspended-level viscometer under conditions set out in ASTM D445. The viscosity of polybutenes increases with the molecular weight of the polymer. Polybutenes are available commercially with viscosity from about 1 to 45,000 cSt at 100°C, corresponding to a molecular weight range from 180 to 6,000. The viscometric properties of a typical range of polybutenes are shown in Tables 3.9 and 3.10.

TABLE 3.9
Physical Properties of Lower-Molecular-Weight Indopol Polybutenes

	Indopol				
Property	**H7**	**H8**	**H15**	**H25**	**H50**
Molecular weight	440	490	570	635	800
Specific gravity, g/cm^3	0.850	0.857	0.865	0.869	0.884
Viscosity at 40°C, cSt	117	—	480	—	—
Viscosity at 100°C, cSt	12	15	30	52	108
Viscosity index	90	—	90	—	—
Pour point, °C	–36	–35	–35	–23	–13
Flash point, COC, °C	>145	>141	>141	>150	>190
Refractive index	1.474	1.478	1.485	1.486	1.490
Polydispersivity index	1.80	1.85	1.85	2.10	1.60

Source: Pathmaster Marketing Ltd., from INEOS published data.
Note: COC = Cleveland open cup.

TABLE 3.10
Physical Properties of Higher-Molecular-Weight Indopol Polybutenes

	Indopol					
Property	**H100**	**H300**	**H1200**	**H2100**	**H6000**	**H18000**
Molecular weight	910	1,300	2,100	2,500	4,200	6,000
Specific gravity, g/cm³	0.893	0.904	0.906	0.912	0.918	0.921
Viscosity at 100°C, cSt	218	630	2,500	4,250	12,200	40,500
Viscosity index	125	173	242	267	306	378
Pour point, °C	−7	3	15	21	35	50
Flash point, COC, °C	>210	>240	>250	>270	>275	>280
Refractive index	1.494	1.497	1.502	1.504	1.505	1.508
Polydispersivity index	1.60	1.65	1.80	1.85	1.80	1.70

Source: Pathmaster Marketing Ltd., from INEOS published data.

Lower-molecular-weight polybutenes have VIs comparable to those of mineral oil base oils. VIs in excess of 170 and up to 380 are available with higher-molecular-weight polybutenes; associated with these grades are relatively high viscosity and pour point. The viscosity of lower-molecular-weight polybutenes shows little dependence on shear rates and could be classed as near-Newtonian fluids. With higher-molecular-weight grades, viscosity begins to show some dependence on shear rate, and at high shear rates, there is evidence to suggest some pseudoplastic behaviour.

It is worth noting that the low-molecular-weight polybutenes have very low pour points, while the high-molecular-weight polybutenes are solids at ambient temperatures. As a consequence, particular attention needs to be paid in a blending plant to the exact grade of polybutene that is required in a specific lubricant formulation.

The flash point of polybutene indicates the proportion of low-molecular-weight polymer in the bulk material. The level of low-molecular-weight polymer present in the polybutene is influenced by the extent of light polymer stripping used by the manufacturer during production. Polybutene grades sold commercially are manufactured to a guaranteed flash point specification. In general, although it need not be the case, flash points increase with the polybutene grade or molecular weight.

Commercially available polybutenes have colours ranging from water white to straw, water contents generally in the region of 20–40 ppm, total acid numbers in the order of 0.02 mg KOH/g and densities in the range of 0.82–0.92.

Each polybutene grade is made up of a distribution of polymer chains of different molecular weights. As one consequence of this distribution, at temperatures in excess of about 80°C polybutenes can lose a proportion of their mass through evaporation of the lowest-molecular-weight polymer molecules. The rate of loss of material decreases with increasing polybutene molecular weight.

Polybutenes differ from most other oils by decomposing at temperatures of 170°C–240°C, through a depolymerisation or unzipping mechanism. The decomposition products are of considerably lower molecular weight than the number-average molecular weight of the grade from which they are derived. Under conditions of rapid depolymerisation it has been shown that the main product of decomposition is isobutylene. Depolymerisation of the polymer chain does not proceed rapidly below temperatures of 180°C–200°C. At these temperatures, it is often difficult to distinguish between evaporation loss of low-molecular-weight polymer molecules and possible loss of polymer fragments through depolymerisation.

The depolymerisation mechanism of polybutenes is very valuable in allowing the higher-molecular-weight grades to volatilise before combustion takes place. This type of decomposition occurs cleanly and completely without the formation of carbon or carry materials and is in contrast to the behaviour of mineral oils, which decompose to leave carbon residues. However, if a polybutene is heated to above its flash point and then ignited, carbon and smoke will be generated as combustion products. In the case of the higher-molecular-weight grades, combustion is not self-sustaining unless the body of the liquid is raised to a temperature at which rapid depolymerisation may also occur.

Polybutene polymers are materials of very low biological activity, as evidenced by their accepted use for many years as components of cosmetics, surgical adhesives and pharmaceutical preparations. The polymers are products of extremely low oral toxicity. Polybutenes are virtually insoluble in water, so their biotoxicity in an aqueous medium is difficult to assess. The International Marine Organisation (IMO) rates polybutenes as unlikely to bioaccumulate and non-hazardous to marine species. In studies based on the European Community (EC) respirometric test for theoretical oxygen demand (ThOD), comparative tests revealed polybutene levels of biodegradation to be less than that for a PAO or mineral oil of similar viscosity. Under the test conditions, none of the materials attained the necessary level of ThOD to be regarded as readily biodegradable as accepted by the Organisation for Economic Co-operation and Development (OECD). Under more favourable conditions, such as higher levels of active sludge acting for a longer period, however, it is clear that biodegradation of the oils including polybutene would be increased. The highly branched structure of the polybutene polymer is thought to be responsible for its resistance to biodegradation.

3.5.4 Polyalkylene Glycols

It is widely known within the lubricants industry that PAG fluids do not exhibit the same characteristics as other lubricant base oils. It is important when assessing the potential use of a PAG that a generic approach is not taken to the performance of the fluid since the flexibility of PAG chemistry means that the performance of the fluid under test has to be assessed on an individual basis. For example, due to polarity increases, a PAG with high ethylene oxide (EO) content will exhibit extremely low mineral and PAO compatibility, but high solubility in water.

PAGs are manufactured using EO, propylene oxide (PO) or (occasionally) higher-molecular-weight alkylene oxides and either water, butanol or a diol as a starting alcohol. As with esters, PAGs are generally manufactured in batches, rather than continuously. The molecular weights of these PAGs range from around 200 to more than 8000, depending on how long the polymerisation reaction is allowed to continue. Typical properties of different types of PAGs are shown in Tables 3.11 through 3.13.

TABLE 3.11
Physical Properties of Butanol-Initiated Ethylene Oxide/Propylene Oxide (50/50) Random Copolymer PAGs

	PAG					
Property	**EO/PO-32**	**EO/PO-46**	**EO/PO-68**	**EO/PO-150**	**EO/PO-220**	**EO/PO-460**
Density at 20°C	1.028	1.036	1.041	1.045	1.050	1.055
KV at 40°C, cSt	32	46	68	150	220	460
KV at 100°C, cSt	7.04	9.9	14.0	29.0	41.0	79.0
Viscosity index	190	210	216	234	241	255
Pour point, °C	–45	–44	–42	–34	–32	–32
Flash point, COC, °C	205	230	230	230	232	235
Refractive index	1.452	1.454	1.457	1.460	1.460	1.460

Source: Pathmaster Marketing Ltd., from Dow Chemical published data.

TABLE 3.12
Physical Properties of Diol-Initiated Ethylene Oxide/Propylene Oxide (60/40) Random Copolymer PAGs

	PAG					
Property	**EO/PO-68**	**EO/PO-150**	**EO/PO-220**	**EO/PO-320**	**EO/PO-460**	**EO/PO-1000**
KV at 40°C, cSt	68	150	220	320	460	1000
KV at 100°C, cSt	12.7	26.4	40.9	68.0	—	—
Viscosity index	193	213	240	250	252	289
Pour point, °C	–40	–38	–38	–36	–35	–31
Cloud point, °C	>100	>100	85	80	75	70
Flash point, COC, °C	220	240	240	245	258	260

Source: Pathmaster Marketing Ltd., from Dow Chemical published data.

TABLE 3.13
Physical Properties of Polypropylene Glycol Monobutyl Ethers

	PAG				
Property	**PO-32**	**PO-46**	**PO-68**	**PO-150**	**PO-220**
Density at 20°C	0.894	0.895	0.896	0.996	0.999
KV at 40°C, cSt	32	46	68	150	220
KV at 100°C, cSt	6.95	9.29	12.9	26.0	37.7
Viscosity index	185	190	193	210	223
Pour point, °C	–42	–41	–35	–32	–31
Flash point, COC, °C	212	215	223	230	230
Refractive index	1.446	1.447	1.449	1.450	1.451

Source: Pathmaster Marketing Ltd., from Dow Chemical published data.

The performance advantages offered by PAGs are well defined and can be summarised as:

- Very high viscosity indices (typically >200).
- Good temperature stability.
- Superior lubricity; reduced wear characteristics and good extreme-pressure performance.
- High hydrolytic stability.
- Good oxidation stability.
- Thermal degradation products are soluble in the base fluid.

Similarly, there are a number of disadvantages commonly reported (perceived or actual) that need to be taken into account when PAG use is considered. These can be summarised as:

- Poor compatibility with mineral and PAO-based fluids.
- Incompatibility with many commonly used paints.
- Poor compatibility with some sealing materials.
- Additive response in some cases.

The compatibility issues associated with PAGs, mineral oils and PAOs are well documented. Essentially, these arise due to the high oxygen content of the PAG increasing the overall polarity of the molecule and thus reducing compatibility. Contamination of a PAG with a mineral oil or PAO (or vice versa) may create two-phase systems and in severe cases gelation. The extent to which contamination becomes a problem is largely a function of the PAG. For example, by incorporating a long-chain alcoholic starter group into the molecule, maximising PO content and incorporating a terminal hydroxyl cap, oil compatible materials can be manufactured.

The compatibility of PAG fluids and alkyd paint has also been well documented. With the advent and use of epoxy-type paint systems, however, occurrences where paint incompatibility occurs have now reduced dramatically. It should be noted, however, that the effect of a PAG fluid on a paint should always be assessed before use.

The aggressiveness of PAG toward elastomers can be regarded as low with most synthetic rubbers. As with paint compatibility, molecular weight and EO content play a significant role in the overall response of an elastomer toward PAG.

A simple comparison of bond strengths (C–C, 84 kcal/mol; C–O, 76 kcal/mol) suggests that the thermal stability of a mineral or PAO-based product would exhibit slightly better performance than PAG. Differential scanning calorimetry (DSC) analysis suggests that a PAG (regardless of oxide structure) will decompose thermally at approximately 250°C, which is comparable to hydrocarbons. Under thermal decomposition PAGs undergo chain scission, producing low-molecular-weight species. These products of degradation are either volatile or soluble within the bulk fluid. Therefore, PAGs, unlike mineral oils or PAOs, do not form sludges or varnish, thus maintaining the cleanliness of the lubricated system.

All ethers are chemically susceptible to oxidative attack at the secondary or tertiary carbon atom adjacent to the oxygen atom. Since PAGs have an oxygen atom at every third position across the polymer chain, it can be assumed that generally PAGs will suffer under oxidative attack. The mechanism for oxidation is similar to that for mineral and PAO-based oils, with the exception that the initial C radical is stabilised by the adjacent oxygen atom.

Due to the ever-decreasing chain length during oxidation, PAG degradation products are volatile, resulting in clean burn-off, an important property in applications such as oven chain lubricants. While the onset of oxidation is stabilised and therefore more likely to occur under oxidising conditions, PAG lubricants are no more likely to oxidise than hydrocarbons. In terms of formulation and final use, PAGs can be successfully inhibited against oxidation via the use of phenolic or aminic antioxidant combinations.

At an average VI of greater than 200, the VIs of PAG-based lubricants are universally known to be the highest of all classes of lubricant base oils. As a result, PAG-based lubricants are capable of operating within extremes of temperature without detrimentally affecting performance. An HVI is an inherent property of PAG, so VIIs are not required when formulating products. Consequently, the shear stability of a PAG is comparably higher than that of many formulated hydrocarbon-based products.

In addition to having an HVI, film-forming properties under extreme pressure and polarity play a major role in the overall lubricity of a fluid. Under extreme pressure, the tendency for a liquid to thicken and hence maintain a lubricating film is measured by the pressure viscosity coefficient (α). The higher the value of α, the greater the likelihood of film maintenance under extreme-pressure conditions.

One factor that plays a major role in the performance of a PAG is polarity. All PAGs exhibit greater levels of polarity than hydrocarbon-based lubricants. This polarity will inevitably increase the affinity of the lubricant to the metal surface, thus potentially increasing boundary lubricity. Additionally, this effect will increase with the inclusion of EO into the polymer backbone. EO is a highly polar molecule that, when polymerised, produces a water-soluble product. Conversely, the pendant methyl groups present in PO reduce the polarity of the polymer to the extent that above a molecular weight of approximately 400 Da, the product becomes almost completely water insoluble.

Random copolymers of EO and PO are generally manufactured via the addition of an oxide premix to the reactor. EO has a faster rate of reaction; thus, the end of the molecule closest to the starter will be EO rich. This concentration gradient and resultant polarity ultimately increase the affinity of the molecule to the metal surface, thus imparting a pseudo-extreme-pressure activity. The depth of this activity is currently unknown; however, as described earlier the performance of this type of material is generally better than that of a polypropylene glycol (PPG) with the same additive system. Issues concerning additive-induced micropitting and bearing damage can therefore be minimised by reducing or even eliminating harmful extreme-pressure and anti-wear additives.

The key attributes of the different types of PAGs include:

- Wide range of viscosities.
- Good anti-wear properties.
- Range of water or oil solubility characteristics.
- Very high viscosity indices.
- Low pour points.
- Good thermal stability.
- Soluble thermal decomposition products.
- Good additive response.
- Non-corrosive to metals.
- Little or no effects on elastomers.
- Low toxicity.

3.6 END-USE MARKETS FOR SYNTHETIC LUBRICANTS

The range of current and potential applications for synthetic lubricants is huge. In order to cover the subject in some systematic way, the applications can be grouped into the following areas:

- Automotive lubricants.
- Compressor oils.
- Turbine and hydraulic oils.
- Gear, circulating and process oils.
- Metalworking fluids.
- Other industrial lubricants.
- Greases.

In total, within these sectors, some 70 end uses for synthetic oils have been identified. A discussion of all the end-use markets for synthetic lubricants is beyond the scope of this book. This is the subject of the book mentioned at the beginning of this chapter.[1]

3.7 CONCLUSIONS

The synthetic lubricants segment of the business remains tough and competitive. It is no longer a pioneering industry, since numerous examples exist of the cost-effectiveness of high-performance oils and greases. Many end-user customers and original equipment manufacturers (OEMs) have accepted synthetic lubricants as having inherently better performances and lower total costs than conventional mineral oils. The lower total costs are often derived from longer equipment operating lifetimes, reduced system maintenance and downtime, higher equipment productivity, improved energy efficiency and reduced wastage and disposal costs.

The market for automotive engine oils has been through a period of unprecedented change in terms of industry specifications, product development and customer expectations at a time of low overall growth in demand for products. This has caused problems of profitability for additive manufacturers and base oil producers, in addition to opportunities for companies with marketing and service support skills. The primary observation is that suppliers with higher-performance products, better marketing and attention to service support are more likely to take market share from suppliers that lack one or more of these customer-driven attributes.

Overall growth in markets for industrial lubricants is likely to continue to be relatively slow for the foreseeable future. World GDP and inflation rates are becoming lower, so increases in demand for industrial lubricants in Asia, South America and Central Europe are likely to be substantially offset by decreases in demand in Western Europe and North America. Quality and performance are likely to be the key to higher prices and product differentiation in industrial lubricants. Unfortunately, higher quality and performance are likely to lead to further increases in product lifetimes, extended drain intervals, lower maintenance and downtime and reductions in lubricant volumes.

However, due to the likely continuing overcapacity in conventional lubricants markets, the market for synthetics is likely to remain competitive. There will be growing inter-product competition, especially in automotive, two-stroke, compressor, bearing, circulation and hydraulic applications. The result is likely to be a constant stream of reformulated, improved and new products based on cost–benefit analyses of customers' needs.

Pricing relationships between synthetic oils and between mineral oils, unconventional base oils and synthetics are now more widely understood, together with product performance advantages or limitations. This has led to more careful and accurate formulation choices, especially in those areas where confusion reigned in the early 1980s. This applies particularly to 0W-XX and 5W-XX engine oils, where PAOs, esters and their competition from Group II, II+ and III base oils are now the preferred blending stocks for advanced formulations with extended drain intervals and low volatility.

Market drivers will continue to favour synthetic and synthetic performance products. It is therefore tempting to forecast that consumption of synthetic oils in all regions is likely to increase at the same rate as it has in Western Europe. The increase in numbers of cars, trucks and buses in countries such as China and India is likely to increase demands for all types of lubricants, but particularly synthetic lubricants.

This is because of greater attention to limiting the emission of greenhouse gases and to enhanced fuel economy.

At the same time, a gradual switch to hybrid vehicles in developed economies, but also in other regions, will put downward pressure on demand for automotive lubricants. It will be interesting to see how extensive the use of all-electric vehicles becomes over the next 20 years.

Unfortunately for manufacturers of synthetic oils, some of them are severely supply constrained. Increases in demand for PAOs are unlikely to be met by increases in supply, due to the constraints imposed by a low increase in demand for linear alphaolefins used in the plastics and detergents industries. Increases in lubricant demand for some esters will need to be balanced with increased demand in the food, chemical and pharmaceutical industries, coupled with raw material supply constraints.

The widening gap between demand for high-performance lubricants and tightness of supply of PAOs and esters is likely to have to be filled by Group II+, III and III+ base oils. Pathmaster Marketing believes that the technical distinctions between "synthetic", "synthetic performance" and "high performance" are likely to become increasingly blurred.

The future for synthetic lubricants remains interesting but not spectacular. Success will come from hard work and the dedication of efforts to meeting real customer needs and solving technical problems.

REFERENCE

1. Rudnick, L.R. (ed.) *Synthetics, Mineral Oils, and Bio-Based Lubricants*, 2nd edition, CRC Press, Boca Raton, FL, 2013.

4 Lubricant Additives *Properties and Characteristics*

4.1 REVIEW OF THE DEVELOPMENT OF LUBRICANT ADDITIVES

Since the invention of the wheel in about 3500 BC, moving machinery has been lubricated by natural oils and fats, because it was found that it was easier to move the parts and they wore out less frequently if some kind of "oily" material was applied to them. Even during the Industrial Revolution (1760–1840), machines were still lubricated by oils and fats of animal or vegetable origin. Although water, and then steam, power was available to drive quite complex machinery, the actual parts to be lubricated were basically very simple, consisting mainly of journal bearings and sliding crossheads. The stresses on the lubricant were generally not very severe in terms of temperature or load, and oiling tended to be both generous and continuous, with considerable losses. Breakdowns which could be attributed to poor lubrication were probably few, although excessive wear in units made of unsuitable materials or badly fitted together was probably quite common. The natural oils and fats contained fatty acids, and their esters, and had inherent boundary lubrication properties.

Toward the end of the nineteenth century, two major events occurred which had a profound effect on each other: large reservoirs of petroleum were discovered and the internal combustion engine was invented. From the work to distil coal to provide fuel and lubricating fractions, it was quickly realised that petroleum represented a major source of both cheaper fuel and lubricants. Since then, developments have been rapid. Petroleum lubricants were produced from the higher-molecular-weight components of crude oils, eventually by vacuum distillation. The clarity, colour and stability of the petroleum oils were improved by various types of refining, including acid treatment, to produce oils which were pale, clear and light stable, and produced no sludge on standing.

Although pure and stable oils were very suitable for use in steam turbines and for electrical insulation, it was found that the oxidation stability of acid refined oils was worse than that of less refined material. It was found that many of the natural inhibiting compounds (particularly sulphur compounds) had been removed in the acid treatment process. Work in Paris and Boston in the 1920s showed that some phenolic and/or amine compounds that were commercially available could inhibit oil oxidation, to produce very stable oils. Phenolic/amine combinations are still used to this day to inhibit turbine oils, compressor oils and other high-quality industrial oils, although the precise compounds may have changed.

Petroleum oils were inexpensive and, if not overrefined, were relatively stable. However, they were found to be lacking in "lubricity" for heavily loaded mechanisms, giving more wear than the natural oils that they replaced. (It would now be described that this was due to a lack of inherent boundary lubrication properties.) The first additives were additions of natural oils to petroleum lubricants, to restore some of this ill-defined lubricity. Later, the natural oils were replaced by purer fatty acids and the synthetic esters derived from them. Similar compounds are found in some industrial oils today.

The internal combustion engine brought major changes to technological development and presented problems of lubrication both for itself and for equipment which rotated faster than steam- or water-powered devices. With rapid development of the automobile, compact self-contained lubrication systems were required, and the concept of an engine sump with splash and/or pumped lubrication was developed. The automobile is a machine which operates intermittently, so the temperature of the lubricant can be high after prolonged use at high power outputs or it can fall to low levels when the vehicle is stored out of use and outdoors. As the use of the automobile developed, it was soon found that there were difficulties in providing lubricants that were sufficiently fluid to enable starting to be achieved in cold weather, but which also provided adequate lubrication (a high enough viscosity) when the engine was being run continuously. The first synthetic additives to be specially developed for motor oils were produced in 1932 by the Standard Oil Development Company (now Exxon Research) to help alleviate this problem. "Paraflow" was a pour point depressant to improve low-temperature flow, and "Paratone" was a polymeric viscosity index (VI) improver based on polyisobutylene to provide increased viscosities at high temperatures.

A major problem with a conventional internal combustion engine, with the connecting rod pivoting at the base of the piston, is that it is not possible to separate the products of combustion entirely from the lubricant. Sufficient harmful material passes the piston rings and enters the crankcase to mix with the lubricant to be harmful both to the oil itself and to the parts of the engine into which it is carried by the process of lubrication and where it forms deposits.

The introduction by the Caterpillar Tractor Company of their new range of high-performance diesel engines for earth-moving equipment was a milestone in engine development in the mid-1930s. More powerful than previous engines and running at higher temperatures, these rapidly ran into problems of piston deposits and piston ring sticking. Caterpillar sought the help of SOCAL and, in 1935, marketed the world's first detergent oil containing aluminium dinaphthenate under their own name. Most oil companies joined a race to develop other detergent diesel oils for marketing under their own brands. One small oil-blending company called Lubrizol was far-sighted enough to hire some research expertise and go on to develop a succession of different additives of increasing utility. They soon decided to give up the oil-blending business to concentrate on additives for motor oils, and have remained a major provider of detergent and other motor oil additives ever since.

The development of detergent additives for diesel engines was greatly accelerated during the Second World War when improved lubricants were required both for military land vehicles and for use in submarines. The impetus continued after the end of the war, and many detergent materials were developed, such as thiophosphonates, phenates, salicylates and sulphonates, the latter available as a by-product from the acid refining of petroleum lubricants.

Nitrogen-containing polymers as dispersants for controlling sludge in gasoline engines were first patented by DuPont in 1954. They were followed by Rohm and Haas with a range of multifunctional compounds, which not only acted as dispersants but also were viscosity improvers and in some cases pour depressants. They were followed in turn by Exxon and other companies with other types of polyester dispersant VI improvers. In the 1960s, Lubrizol developed the succinimide dispersants, which were both more efficient and capable of being developed for use in diesel engines, where the polyester types were unsuitable. The use of major amounts of dispersants in diesel oils, however, did not come until it was found possible to formulate oils which would satisfy the severest requirements of both gasoline and diesel engine lubrication.

The use of detergents in gasoline engine oils was slow to appear because many people believed introducing detergent oils into engines which had previously operated on non-detergent types could bring problems of oil-way blockage due to haphazard sludge removal. More importantly, it was found that the use of detergents in the higher-revving and lighter-built gasoline engine produced wear in the valve mechanisms which had not been seen in diesel engines. Gasoline engines have very highly stressed valve gear, and detergent additives tend to displace from the metal surfaces the boundary layer and natural compounds in the oil which provide essential lubrication. The answer lay in the use of zinc dialkyldithiophosphate (ZDDP) in gasoline formulations as an anti-wear additive to counteract the effect of the detergents. ZDDP was initially developed in 1941 by Lubrizol as a bearing passivator and oil antioxidant, but at higher treat levels it enabled detergent gasoline oils to be developed, which soon proved to be instrumental in prolonging the overall life of motor car engines. Not only do the detergents reduce deposit formation, but also their alkalinity neutralises fuel-derived acids which cause bore wear.

The period between the late 1960s and the early 1970s was the last period of fundamental change in additive technology, from which time the developments have been more evolutionary than revolutionary. During this period, the concept of multipurpose gasoline and diesel lubricants became thoroughly accepted, and the use of barium-based detergents began to be phased out because of toxicity fears, while new magnesium-based detergents took their place. New polyolefin viscosity improvers were developed which were suitable for diesel use, and this again helped to further the case for the multipurpose gasoline or diesel lubricant.

Low-friction oils were introduced in the 1980s as a contribution to fuel saving, using variations on boundary lubrication technology. Recent developments have been mainly concerned with meeting increasingly severe requirements for detergency, dispersancy, wear prevention and oxidation control at reasonable cost by optimisation of additive treatment. Restrictions on phosphorus levels to prevent exhaust catalyst poisoning and ash levels to reduce deposits have added to the difficulties of the formulator.

The identification of oil volatility as a cause of engine deposits, as well as high oil consumption, has led to incorporation of special synthetic base oils into motor oils.

These are in general clean burning and aid the formulator, but they lead to formulation balances different than those for conventional base oils.

4.2 FUNCTIONS OF LUBRICANT ADDITIVES

Additives can either enhance the existing properties of a basic lubricating oil or impart properties which the raw base oil does not possess. Additives can provide *physical* improvements in the oil quality, for example, in viscosity and in low-temperature flow, but many additives are *chemical* in nature and serve either to maintain the oil itself in good condition or to assist in the process of maintaining the mechanism in a satisfactory state while minimising wear and deposit formation.

Lubricant additives most usually have one or more of four basic functions:

- Enhance existing base oil properties: Anti-wear, lubricity, corrosion inhibition.
- Impart new properties: Detergency, emulsifiability.
- Provide physical improvements: VI, pour point, anti-foam.
- Provide chemical properties: Oxidation stability, extreme pressure (EP), load carrying.

The choice of additives in a lubricant is dependent on the use to which the oil is to be put, so it is necessary to consider the major types of lubricants separately. The largest additive usage is in motor oils (crankcase oils), but many types of additives are used in a wide range of lubricant types.

4.3 TYPES OF LUBRICANT ADDITIVES

Although lubricant additives are chemicals, they are grouped according to the function they perform in an oil or grease. The main types of additives are therefore classified as:

- VI improvers.
- Detergents.
- Dispersants.
- Pour point depressants.
- Oxidation inhibitors.
- Anti-wear additives.
- EP additives.
- Corrosion inhibitors.
- Metal passivators.
- Demulsifiers.
- Emulsifiers.
- Friction modifiers.
- Anti-foam additives.
- Biocides.

- Anti-mist additives.
- Tackiness agents.
- Anti-fatigue additives.
- Removability agents.
- Scourability agents.
- Dyes.

We discuss the functions of each of these types of additives briefly, together with examples of the chemicals that are able to provide these functions in a blended lubricant.

Some additives have multifunctional properties; they can do more than one action in a lubricant. For example, many ZDDPs and their additives function as antioxidants as well as anti-wear additives. Other additives have directly opposing functions, meaning they are mutually incompatible in a lubricant formulation. For example, emulsifying additives work to suspend droplets of water in oil (or droplets of oil in water), while demulsifying additives work to displace droplets of water out of the oil phase as quickly as possible. A lubricant formulator would never use an emulsifier in the same lubricant as a demulsifier.

A detailed description of all the modes of action of these additives is beyond the scope of this book; it is the subject of a separate book.[1] What is most important for blenders of finished lubricants are the properties and characteristics of each additive.

4.3.1 Viscosity Index Improvers

The VI of an oil is an empirical number that provides a measure of the decrease in viscosity of the oil as the temperature increases. The viscosity of all lubricating oils decreases as the temperature is raised, but an oil which is relatively thinner at low temperatures and relatively thicker at higher temperatures is said to have a high VI. This property of a base oil can be controlled by refining but is often improved by the addition of a VI improver, sometimes called a viscosity modifier (VM).

These are various types of polymers which are only moderately soluble in oil at low temperatures, when they exist as tightly coiled individual chains which have little effect on the oil viscosity. At higher temperatures, the polymers are more soluble and the chains become "solvated" and open out into the body of the oil. Mutual interference and a greater effective size gives an apparent increase to the viscosity of the oil, an effect which is seen by an engine provided that the oil flow is not so rapid that "temporary viscosity loss under shear" can take place. (This is a reversible process, to be distinguished from the shearing of polymer chains in devices such as gearboxes or under high-speed, high-temperature driving. When the polymer chains are physically sheared by such means, the viscosity reduction is permanent.)

The first VI improver was polyisobutylene, a polyolefin material not dissimilar to the modern types, such as the olefin copolymers and polyisoprene. For a long period between the late 1930s and the 1970s more popular types were based on polyesters of various sorts, which were both easy to manufacture and could also readily be tailored to provide both pour point depressancy and some dispersancy in the same product. In general, the use of polyesters has declined because their thermal stability is not adequate to permit their use in diesel engines, where they tend to produce piston deposits.

Newer types of VI improvers include polymethyl methacrylates (PMAs) and olefin copolymers (OCPs). Most suppliers of these additives market a range of molecular weight products, so that lubricant formulators can balance or optimise the performance properties of different types of lubricants.

4.3.2 Detergents

In an internal combustion engine, fuel is burned to produce power and the main combustion products are water and carbon dioxide. However, fuel contains substances other than pure hydrocarbons, and sulphur in diesel fuel and the scavengers in leaded gasoline produce strong mineral acids. These combine with the water of combustion, and while most pass out of the exhaust, a proportion enter the crankcase as "blow-by" past the piston rings and mix with the oil. This can lead to rapid corrosive wear of the engine, and also catalyses deposit formation from partially burned fuel residues (which are also present in the blow-by) and promotes oil degradation and consequent further deposit formation.

Alkaline additives (detergents) are added to the oil to neutralise such acids. Metal sulphonates, particularly the highly overbased varieties, are excellent neutralisers of acids but tend to promote oil oxidation. Overbased phenates and salicilates, and the earlier thiophosphonates, act as inhibitors as well as neutralisers for the acids, and are therefore bifunctional detergent inhibitors. Other types of detergents include metal naphthenates and metal thiophosphonates, where the metals are usually calcium, sodium or magnesium. (Barium was used in the past, but its general toxicity in compounds means that it has been phased out gradually.) In practice, an optimum blend of highly neutralising sulphonates with a phenate or salicylate is often used in current formulations. Dispersants (see below) also assist in the prevention of deposits by isolating and suspending solid and liquid contaminants, and are sometimes known as "dispersant detergents".

4.3.3 Dispersants

Particularly in the case of diesel engines, very fine particles of unburned fuel, known as soot, enter the crankcase in the blow-by past the piston rings. Soot in engine oil can cause excessive wear and can lead to the formation of sludge. Sulphonates are quite good dispersants for diesel sludge and were initially the principle means of suspending soot between oil changes.

Gasoline engines generally operate at lower temperatures than diesel engines and therefore retain higher water contents in the oil. This leads to the formation of a different type of sludge. Free water can also separate out in the cooler parts of the engine. Gasoline dispersants were initially nitrogen-containing polyesters which were often also used as viscosity improvers. These were not very thermally stable and were therefore unsuitable for diesel engines. The "succinimide" dispersants made from polyisobutylene, maleic anhydride and a polyamine are superior products for dispersing wet sludge in gasoline engines. They were also able to be developed to a sufficient level of thermal stability to be useful in diesel oil, in which they now form a significant part of the additive treatment. Pentaerythritol is an alternative polar constituent to polyamines.

Dispersants generally function through having a polar part of the molecule which attaches itself to non-oily impurities (water, soot and other combustion products) and an oil-soluble hydrocarbon chain which suspends the impurity particles individually within the body of the oil. The sludge is thereby removed from the engine at times of oil changes or due to normal oil consumption, either from burning or from leakage losses. Isolated and suspended in oil, the contaminants have little adverse effect on the engine.

The main types of dispersants used in either gasoline or diesel engine oils include polyisobutene succinimides, polyisobutene succinate esters, polypropylene amine carboxylates, polypropylene alkyl phenol amines, alkyl sulphonates, alekenyl phosphonic acid derivatives and polyamin imidazolones.

4.3.4 Pour Point Depressants

Pour point depressants provide good low-temperature flow for cold weather use and aid starting in engines. They permit a given level of pour point to be achieved more economically than by heavily dewaxing the base oil. The structure of a pour depressant contains a carbon chain which crystallises with the wax in the oil, but also contains large side chains or chemical groups which interfere with wax crystal formation and prevent large wax crystals forming and interlocking to "gel" the oil. Pour depressants do not prevent the deposition of wax from an oil on cooling, but change the nature of the wax into a form which is not harmful.

The most commonly used pour point depressants include PMAs, alkylated naphthalene polymers and alkylated phenol polymers. The PMAs used as pour point depressants have molecular weights that are generally higher than those PMAs used as VI improvers.

4.3.5 Antioxidants (Oxidation Inhibitors)

An antioxidant is a compound that is effective in preventing oxidation caused by molecular oxygen. Such additives are used widely in the food, food products, rubber and plastics industries, in addition to their use in the oil industry. They were used in foods long before their function was appreciated. Spices, such as sage, cloves, oregano, rosemary and thyme, were used as stabilisers and taste enhancers for several centuries before it was established that they contain natural phenolic compounds that prevent the development of peroxides and increase the stability of fats to oxidation.

In lubricating oils, commonly used antioxidants include phenols, cresols, aromatic amines, phenothaiazines and sulphur compounds. Metal salicylates and phenothazines are sometimes used as oxidation inhibitors. Many types of ZDDP anti-wear additives (see below) are also powerful antioxidants, so motor oils and hydraulic oils often contain ZDDPs that function as combined anti-wear and antioxidant additives.

4.3.6 Anti-Wear Additives

In modern engines, particularly gasoline engines, high loads and high sliding speeds pose severe lubrication problems to components such as the inlet and exhaust valves. In these situations, simple boundary lubricant additives are not sufficiently effective, so higher anti-wear performance is required. This also applies to hydraulic systems and some types of compressors. One particularly effective type of anti-wear additive is ZDDP. This family of additives, introduced by Lubrizol in 1941, acts successively as a boundary lubricant additive, a mild EP additive or a severe EP additive, depending on the conditions and stress on the lubricant. They also possess particularly good antioxidant properties. Differences in the hydrocarbon side chains in the various types of ZDDPs confer different levels of thermal stability and anti-wear performance. There are many possible variations which can be used, but in general terms, the effect of the side chains is:

- Secondary alkyl: Poorest stability, greatest anti-wear performance.
- Primary alkyl: Intermediate performance.
- Aryl (aromatic): Best stability, delayed anti-wear.

Other types of anti-wear additives include organic phosphates, organic phosphites, phosphate esters, polyethylene glycol esters, polymer esters, dibasic acid esters and fatty acids. Molybdenum dialkyldithiophosphates are also now being used as anti-wear additives.

4.3.7 Lubricity Additives

Polar lubricant additives, sometimes called "lubricity" or "oiliness" additives, have long carbon chains with one or more ionically charged groups at one end. These ionically charged, or polar, groups adhere strongly to metal surfaces in such a way that the long carbon chain sticks up approximately at right angles to the surface. When there is a sufficiently high concentration of these additives, they form a film on the metal surface, with the polar groups packed on the surface and all the long chains oriented away from the surface. Although the long chains are not rigid; they form a deformable film several molecules thick that minimises metal-to-metal contact and significantly reduces friction. They also retard the plastic flow of asperities (metal surfaces are never flat at microscopic scales and the peaks on the surface are called asperities) and thus minimise galling or welding.

In some types of industrial lubricants, particularly metalworking fluids, examples of these lubricity additives include animal and vegetable fats and fatty acids, fatty acid soaps, long-chain carboxylic acids, polyalkylene glycol esters and polymer esters. These additives are also used as friction reducers and friction modifiers in several types of lubricants. They also function as wetting agents on inactive metal surfaces and as load-carrying additives on active surfaces where active EP additives cannot be used due to the possibility of staining.

Older formulations of some metalworking lubricants used animal fats, such as tallow and lard, in preference to vegetable oils, which used to be expensive and difficult to obtain. However, animal fats tend to degrade oxidatively and thermally much faster than vegetable fats, particularly palm oil and rapeseed oil. Now that vegetable oils are widely available (for the manufacture of margarine and other cooking ingredients), they have almost entirely replaced animal fats.

4.3.8 Extreme-Pressure Additives

Boundary lubricants fail to provide adequate protection of metal surfaces when temperatures become high, since the surface film melts or decomposes. In order to protect metal surfaces in machines and operations when high and localised pressures and temperatures exist, EP additives are used in higher-performance pressing and drawing lubricants.

The majority of EP additives are of the chemical type, being sulphurised, chlorinated or phosphorised organic compounds that react with metal surfaces to form a firmly bonded film of metal sulphide, chloride or phosphate. These films are good lubricants in their own right and have lower shear limits than localised, tiny, metal-to-metal welds. The metal sulphide, chloride or phosphate films are constantly being broken and reformed as the metal is being formed.

Examples of EP additives include sulphurised fats and fatty acids, chlorinated hydrocarbons and organic phosphates. These materials cannot be used on active metal surfaces, particularly copper, brass, magnesium alloys and silver, due to the high likelihood of staining. Additionally, some short- and medium-chain chlorinated hydrocarbons are being phased out in many lubricants, due to their environmental toxicity. Legislation in many countries now limits the chlorine content of many lubricants to a few parts per million.

4.3.9 Bearing Corrosion Inhibitors

When higher loads on bearings led to babbit metal (tin/copper) being replaced by harder materials such as copper, lead and bronzes (copper/tin), problems of attack by organic acids on the bearings were experienced. The attack arose from oil degradation. Various compounds were found to reduce this problem, but some sulphur compounds were found to be more effective, because they reduced oil degradation, by acting as antioxidants, and inhibited the surface against attack. Modern oils, which need to have high levels of corrosion and oxidation inhibition for other reasons, do not normally require special treatment.

4.3.10 Corrosion Inhibitors or Anti-Rust Additives

In gasoline engines running in colder environments, water can condense in the cooler parts of the engine, such as under the rocker cover or the timing case. To prevent corrosion (rusting) of ferrous components, the oil needs to have specific anti-corrosion properties. Corrosion inhibitors function by coating the metal surface with a hydrophobic layer of additive which prevents water access. Sulphonates, also used as detergents, are effective, but their action can be reinforced if necessary by certain organic acids (such as succinnic acid) or their salts.

Because *rust* is a term usually associated with iron and steel surfaces, corrosion inhibitors are most usually added to lubricants that are used in machines that are made primarily from ferrous metals. Most motor oils, gear oils, hydraulic oils, compressor oils, turbine oils and many metalworking fluids (both neat and water miscible) will contain corrosion inhibitors.

Examples of corrosion inhibitors include amine borates, amine dicarboxylates, alkyl sulphonates, calcium sulphonates, alkanolamides, carboxylic acids, sulphonamido carboxylic acids, alkyl sarcosines, amides and polyalkoxy phosphates.

4.3.11 Metal Passivators

These additives are similar to corrosion inhibitors but are used most often in lubricants that come into contact with non-ferrous metals, notably copper, brasses, bronzes, magnesium alloys, some aluminium alloys and silver. Their mode of action is similar to corrosion inhibitors, but their purpose is to prevent or minimise attack of the metal surface from aggressive sulphur or nitrogen compounds present in the lubricant. They are generally required only in some types of gear oils, turbine oils, hydraulic fluids and metalworking fluids. The most common types of metal passivators are benzotriazoles, thiadiazoles, thiadiazoles, carbamates mercaptobenzothiazoles and dimercaptothiadiazole and derivatives.

4.3.12 Demulsifiers (Anti-Emulsion Additives)

In the cooler parts of a gasoline or diesel engine the water-in-oil emulsion produced by the action of the dispersant on the water formed during combustion can invert to a white oil-in-water mayonnaise. Secondary emulsifiers can prevent or reverse this emulsion sludge formation, being either incorporated in the dispersant or added separately. Because they have the effect of preventing or dispersing the white emulsion sludge, these compounds are often referred to as demulsifiers. This description is correctly applied to an alternative type of additive which "splits" the emulsion into its constituents.

Droplets of free water in gear oils, hydraulic oils, turbine oils and compressor oils are highly likely to seriously impact the anti-wear, oxidation stability, thermal stability and corrosion-inhibiting performance of these lubricants. Although rapid separation of free water from these oils is normally achieved by careful refining of the base oils, the use of demulsifiers can help in arduous applications.

The most commonly used demulsifiers include anionic alkoxylated alkyl phenol resins, polyalkylene glycol block copolymers, block copolymers of ethylene or propylene oxides with glycerol, polyamines or polyols and high-molecular-weight calcium or magnesium sulphonates.

4.3.13 Emulsifiers

Some types of lubricants, particularly metalworking fluids, need to function as both coolants and lubricants. Reduction of tool, chip or workpiece temperatures can be achieved both by conducting the heat away from the cutting zone and by lubrication the chip–tool interface. Many metalworking fluids must have a combination of high thermal conductivity, high specific heat capacity and good lubricating ability.

Water has excellent thermal conductivity and high specific heat capacity but is a very poor lubricant and is corrosive to ferrous metals. Water is not effective in reducing friction between the chip and tool face and can contaminate the lubricating oils used on the sliding and rotating surfaces of the machine, thus reducing the smoothness of running and increasing wear. Mineral oils are excellent lubricants but have relatively low thermal conductivities and specific heat capacities. They have the required characteristics for lubricating the chip–tool interface and are compatible with the other lubricating oils used in the machine. Unfortunately, their thermal conductivities and specific heat capacities cannot be improved by the use of additives, although the use of a low-viscosity oil allows greater flow of fluid over the tool and can give better dissipation of heat.

Relatively stable combinations of water and oil are most usually achieved by means of emulsifiers, additives that encourage the suspension of one liquid in another, immiscible, liquid. Such compounds are used in the preparation of some foods (such as margarine and ice cream), cosmetics, lotions and certain pharmaceutical creams. In "soluble-oil" metalworking fluid emulsions, the oil does not in fact dissolve in water. They are essentially mineral oils blended with emulsifiers and other additives, so that when the blend is added to water and stirred a dispersion of very small oil droplets in a continuous phase of water is produced, forming an oil-in-water emulsion.

Both anionic and non-ionic emulsifiers are used in water-miscible fluids. The choice of emulsifier is normally the critical area of formulation development. Anionic emulsifiers are the fatty acid soaps, glycol esters and sulphonates. Non-ionic emulsifiers are likely to be straight or branched chain organic chemicals, typically nonyl phenol ethoxylates of differing molecular weights.

Commonly used emulsifiers include alkyl sulphonates, alkyl-aryl sulphonates, fatty acid soaps, alkanolamides, glycerol mono-oleate, sorbitan mono-oleate, fatty acid ethoxylates, fatty alcohol ethoxylates, alkyl phenol ethoxylates and polyethylene glycol esters.

4.3.14 Friction Modifiers

The demand for energy-saving performance in many lubricants has led to new developments of old boundary lubricant technologies. Additives that produce small but measurable reductions in friction are now being used to improve energy efficiencies in engine oils, gear oils and automatic transmission fluids. These additives are long-chain molecules with a polar head which attaches itself to the metal surfaces and prevents metal-to-metal contact under boundary conditions.

It is worth noting here that the biggest contribution to energy efficiency with lubricants has been the increasing use, worldwide, of lower-viscosity and synthetic base oils. This has led to lower-viscosity engine oils, gear oils and transmission fluids. The trend was started in Western Europe in the late 1980s and has progressed, first in North America, then Eastern Europe, then Asia and now everywhere. For example, the use of 20W-50 viscosity-grade gasoline engine oils declined, to be increasingly replaced by 15W-40 viscosity grades, then 10W-30 viscosity grades, then 5W-30 viscosity grades and lastly 0W-20 viscosity grades. In the future, 0W-16 and 0W-12 viscosity grades will start to become available, for use in hybrid gasoline–electric vehicles. The ability to use lower-viscosity lubricants depends on many factors, but the choice of base oils, viscosity improvers, anti-wear additives and friction modifiers can all be important.

Typical friction modifiers used in lubricants are high-molecular-weight fatty acids, linear alkyl phosphites and phosphate esters, phosphoric and thiophosphoric acids, long-chain carboxylic acids, esters, ethers, amines, amides and imides, organic fatty alcohols, amides and esters, complex esters and methacrylates.

4.3.15 Anti-Foam Additives (Foam Suppressants)

Engine oils that are highly additive treated, particularly those with high sulphonate contents, can foam in the crankcase, which leads to oil loss and/or poor lubricating ability. Small quantities of silicone oil suppress foam formation and are normally added to engine oils. Silicones are often added in the additive manufacturing process, in the production of the additive packages, in the blending of the final lubricant or in all three cases. It is, however, important that excessive amounts of silicone are not added as this can reverse the process of foam suppression and also leads to air entrainment.

Formation of excessive foam in gear oils, hydraulic oils, turbine oils and compressor oils also causes significant problems. Foam is not a good lubricant and can impact oxidation and corrosion inhibition properties. In addition to silicone anti-foam additives (only used in parts per million amounts), other anti-foam additives include high-molecular-weight polyalkylene glycols, polyethers and alkoxylated aliphatic acids.

4.3.16 Biocides

Biocides are chemicals that kill microorganisms, such as bacteria and fungi. They are added to water-miscible metalworking fluids in small amounts to help control the contamination of the fluids by ever-present microbes. However, the use of some biocides has been coming under increasing pressure for perceived environmental and toxicological reasons. Also, some microorganisms are developing resistance to some biocides and even combinations of biocides, and this process is likely to continue as microbes evolve and continually adapt to their changing environments.

4.4 ADDITIVE PACKAGES

Several manufacturers of additives, particularly Lubrizol, Afton, Infineum and Oronite, provide pre-prepared combinations of additives, known as "additive packages". These packages make blending finished lubricants much easier, as they are always supplied as liquids.

An additive package may contain a mixture of anti-wear, oxidation inhibitor, corrosion inhibitor, detergent and dispersant additives. For automotive engine oils, an additive package that contains detergent, dispersant, oxidation inhibitor and anti-wear additives, dissolved in a "diluent" oil, is known in the lubricants industry as a "detergent inhibitor" or "DI pack".

Additive packages are available commercially for:

- Automotive and industrial engine oils, for both gasoline light-duty diesel engines or heavy-duty diesel engines.
- Automotive and industrial gear oils.
- Automatic transmission fluids.
- Hydraulic fluids.
- Compressor oils.
- Turbine oils.
- Metalworking fluids.

4.5 PROPERTIES OF LUBRICANT ADDITIVES

Lubricant additives are manufactured and sold in a huge range of physical conditions. Some are low-viscosity liquids, some are high-viscosity liquids and yet others are solids, generally powders, flakes, pellets or beads.

Example physical and, for some additives, chemical properties of a number of categories of additives are shown in Tables 4.1 through 4.6. Showing physical and chemical properties for any more than a selection of additives would be exceedingly difficult and time-consuming; most formulators of lubricants have several thousand different additives from which to choose. Pathmaster Marketing has attempted to show examples of the full range of different physical and chemical properties of additives.

TABLE 4.1
Typical Properties of Selected Viscosity Index Improvers

	Viscosity Index Improver				
Property	**A**	**B**	**C**	**D**	**E**
Colour, ASTM D1500	<0.5	<0.5	<0.5	<0.5	<1.0
Density	0.94	0.95	0.94	0.94	0.94
KV at 100°C, cSt	674	838	882	1064	1100
KV at 40°C, cSt	—	—	—	—	—
Viscosity index	—	—	—	—	—
Pour point, °C	—	—	—	—	—
Flash point, °C	150	130	100	140	140
Shear stability index, DIN 51382, 250 passes	4	0	10	4	10

Source: Pathmaster Marketing Ltd., from numerous manufacturers' publications.

TABLE 4.2
Typical Properties of Selected Oxidation Inhibitors

	Oxidation Inhibitor				
Property	**A**	**B**	**C**	**D**	**E**
Colour, ASTM D1500	—	—	—	—	—
Density	0.96	0.98	0.98	1.02	1.15
KV at 100°C, cSt	—	—	—	—	—
KV at 40°C, cSt	120	280	300	Solid	Solid
Viscosity index	—	—	—	—	—
Pour point, °C	—	—	—	—	—
Flash point, °C	152	185	123	273	297

Source: Pathmaster Marketing Ltd., from numerous manufacturers' publications.

TABLE 4.3
Typical Properties of Selected Extreme-Pressure Additives

	EP Additive				
Property	**A**	**B**	**C**	**D**	**E**
Colour, ASTM D1500	8.0	4.5	8.0	5.0	5.0
Density	—	—	—	—	—
KV at 100°C, cSt	—	—	—	—	—
KV at 40°C, cSt	25	30	50	250	350
Viscosity index	—	—	—	—	—
Pour point, °C	9	−10	9	−10	−10
Flash point, °C	196	—	196	180	180
Four-ball weld load, ASTM D2783, kg	250	315	315	240	400

Source: Pathmaster Marketing Ltd., from numerous manufacturers' publications.

Each lubricant blending plant needs to have a detailed description of the physical and chemical properties of all the additives used in the lubricants it blends, for four reasons:

- To check each batch of purchased additive against the specification for that additive.
- To understand the properties and characteristics to enable blending conditions and equipment to be tailored to suit all the additives in each blend.
- For health and safety protection of employees.
- For environmental compliance of blending, handling, packaging and transportation operations and accidental spillage.

TABLE 4.4
Typical Properties of Selected Corrosion Inhibitors

	Corrosion Inhibitor				
Property	**A**	**B**	**C**	**D**	**E**
Colour, ASTM D1500	—	—	—	—	—
Density	1.15	0.96	1.02	1.03	1.52
KV at 100°C, cSt	—	—	—	—	—
KV at 40°C, cSt	80	350	850	1750	Solid
Viscosity index	—	—	—	—	—
Pour point, °C	—	—	—	—	—
Flash point, °C	na	>130	130	130	na

Source: Pathmaster Marketing Ltd., from numerous manufacturers' publications.
Note: na = not applicable.

TABLE 4.5
Kinematic Viscosities of Selected Additive Types

Additive	Kinematic Viscosity at 40°C, cSt
Sulphonate A	300
Sulphonate B	2000
Sulphonate C	43
Succinic acid derivative A	900
Succinic acid derivative B	1100
Fatty acid ester A	30
Fatty acid ester B	45
Fatty acid ester C	48
Olefin A	10
Olefin B	50
Fatty ester + olefin A	640
Fatty ester + olefin B	750
Fatty acid	500

Source: Pathmaster Marketing Ltd., from numerous manufacturers' publications.

Most additives used in lubricating oils are supplied as liquids, since this makes pumping and mixing much easier. Solid additives need to be dissolved in base oil either before or during blending. Many additives are unstable at high temperatures (>100°C). Some additives have very light colour; others are very dark in colour.

The data shown in Tables 4.1 through 4.6 indicate that different additives, from different manufacturers, in the same group of additives (for example, anti-wear additives) often have very different physical and/or chemical properties. The performance

TABLE 4.6
Elemental Contents of Selected Anti-Wear Additives

	Element, % wt				
Additive	**Zn**	**P**	**S**	**N**	**Oil Content, % wt**
ZDDP A	9	17	8.5	—	20
ZDDP B	10.5	19	9.5	—	10
ZDDP C	8	15	7.5	—	10
ZDDP D	9.5	16	8	—	—
Ashless anti-wear A	—	8	8	—	—
Ashless anti-wear B	—	5	—	3	—
Ashless anti-wear C	—	3	7	1	30

Source: Pathmaster Marketing Ltd., from numerous manufacturers' publications.
Note: Zn = zinc; P = phosphorous; S = sulphur; N = nitrogen.

improvements they provide in blended lubricants may be very similar, but the properties can be very different.

This means that, in a blending plant, simply swapping one additive for another one in a specific lubricant formulation is usually not possible and, in fact, may have very damaging consequences for the physical and/or chemical properties of the finished product. The blended lubricant may not meet the specifications for which it is intended in terms of its viscosity, pour point, flash point, corrosion resistance, elemental contents or other important properties.

Similarly, the additive packages available from additive manufacturers may have different physical properties, so may not be directly interchangeable. An illustrative example of a modern DI pack for a gasoline engine oil formulation is shown in Table 4.7.

Again, although the performance properties of a finished blended lubricant may be very similar (or nearly identical) for additive packages supplied by different additive manufacturers, the chemical compositions of those packages may be very different. Manufacturers of additive packages spend a great deal of time and money to optimise the products they market. This will become more evident in Chapter 5.

4.6 EFFECTS OF ADDITIVE PROPERTIES ON LUBRICANT BLENDING

Additives that have low viscosities are much easier to mix with base oils, so blending times can be quite short. Additives that have high viscosities are much more difficult to mix with base oils and can require careful addition to the blend to minimise the risk of layering in the blending vessel.

Solid additives are easier to blend when premixed with base oil and then added to the blending vessel as a liquid.

TABLE 4.7
Illustrative DI Pack for a Gasoline Engine Oil Formulation

Component	% wt
Dispersant	60
Sulphonate	16
Phenate	7
ZDDP	8
Phenolic antioxidant	4
Amine antioxidant	5
Total	100

Source: Pathmaster Marketing Ltd.

Additives that are susceptible to oxidation, particularly oxidation inhibitors and anti-wear additives, should not be mixed by air blowing. Additives that are thermally unstable at high temperatures, particularly EP and anti-wear additives, must be heated and blended very carefully; otherwise, the additive may be degraded during blending. If this occurs, the finished lubricant may not have the performance properties required by a customer.

Additives that may cause foaming, such as detergents, dispersants and some corrosion inhibitors, should not be blended using air blowing or vigorous agitation. Additives that have comparatively low flash points cannot be heated much above 60°C.

Blending is easier and faster to achieve when the viscosity of the mixture is lower, but components should not be heated to a bulk temperature above 60°C; otherwise, oxidation may start to cause problems.

Additives in drums can be preheated in a dedicated hot room, to lower their viscosities and make emptying drums and blending easier. Steam coils in storage tanks and blending vessels allow much better control of both heating rate and temperature than electrical heaters. Silverson mixers tend to be more efficient than paddle mixers in blending vessels, but beware of foaming.

Another very important factor that needs to be considered by blending plant managers, supervisors and operators is the competition and interaction between the various types of surface-active additives.

Anti-wear, lubricity, EP, corrosion inhibitor, metal passivator and friction modifier additives all form layers on metal surfaces. Whichever one gets to the surface first forms the strongest layer and has the biggest effect. Lubricant formulations need to balance the types and amounts of each of these additives in order to achieve the desired lubricant performance properties. For example, too much anti-wear additive is likely to affect the corrosion protection or friction modification performance of the lubricant. Too much metal passivator is likely to affect the anti-wear and/or corrosion-inhibiting properties of the lubricant.

It is therefore critically important when blending a lubricant that the precise formulation is blended, with accuracy of the highest priority. How to achieve this is discussed in Chapters 6 and 7. The reasons why it is so important are presented in Chapter 5.

4.7 SUMMARY

A huge range of additives, with very many properties and characteristics, are used in formulating and blending finished lubricants. Some additives are easy to blend, while others are not. Care must be taken with each lubricant type and each blend; blending different lubricants does not involve a standard procedure.

REFERENCE

1. Rudnick L.R. (ed.) *Lubricant Additives: Chemistry and Applications*, 2nd edition, CRC Press, Boca Raton, FL, 2009.

5 Lubricant Formulation and Ease of Blending

5.1 INTRODUCTION

Before a lubricant can be manufactured in a blending plant, its formulation (the types and amounts of each component) needs to be determined, evaluated and tested by a development chemist in a research laboratory. This is part of the business planning and marketing activities of the lubricant supplier.

The process of selecting which base oil (or oils) and which additives to use is complex and time-consuming, usually relying on a great deal of previous experience coupled with much trial and error. Formulation chemists are not particularly clever or diligent; they succeed by standing on the shoulders of all those who went before them.

However, formulation chemists also need to understand whether the formulation they are developing can be manufactured cost-effectively in a blending plant. Equally, blending plant staff need to understand why the formulations they are required to blend can, on occasions, be so complicated.

5.2 THE NEW PRODUCT DEVELOPMENT PROCESS

5.2.1 Idea Generation

Most manufacturers of products and services suffer not from a shortage of ideas but from the challenge of separating the good ideas from the bad. Any initial product idea should be aimed at solving a specific problem or providing defined customer benefits, and the idea should support and enhance the company's overall strategic mission.

Ideas for new products come from various sources, both inside and outside a company. Within a company, the most likely sources of new product ideas are sales, marketing, manufacturing, new product development groups and management, although any member of staff has the potential to suggest a good idea. Outside sources include distributors, independent researchers, competitors, government agencies and existing as well as potential customers. The steps in going from an idea to a commercial product are illustrated in Figure 5.1.

During the last 30 years, idea generation and innovation within a company has become known as "intrapreneuring". It involves encouraging and enabling employees within an organisation to develop new product ideas and see them through to profitability. Companies encourage intrapreneuring by allowing staff to select themselves as intrapreneurs, not requiring them to turn their projects over to others and allowing them to make major decisions personally. Intrapreneurs are

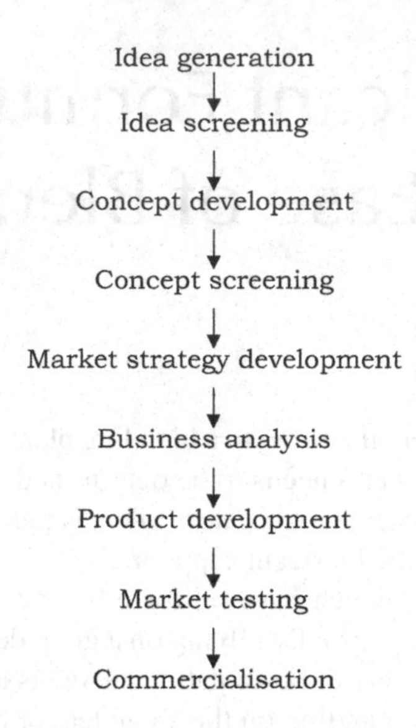

FIGURE 5.1 New product development process.

able to use resources as they choose, even those from outside the company; are given time and money to pursue their product developments; and are allowed to take risks and make mistakes in the process. They are encouraged to make small accomplishments and are not required to achieve sizeable gains, and they are allowed to work in small teams composed of people throughout the organisation. Because of these numerous freedoms, intrapreneuring violates many bureaucratic principles and practices. Therefore, without senior management support, it will have little hope for success.

Successful new product ideas are focused on satisfying some particular customer need. Customers should usually be in the best position to recognise a need and should be the first to request a solution. Research studies, however, indicate that some companies use customer ideas in developing new products, while others do not. A possible explanation is that it takes a long while to develop new products from original ideas, so that by the time many new products have been developed, the source of the original idea has been forgotten.

New product development can be inspired by technological change as well as customer need. However, product development should not lead to companies emphasising either a technological focus or a customer focus. The emphasis should be placed on achieving the proper coordination between the two areas.

An effective research and development (R&D) programme must be based on several directing inputs. First, it should be concerned with satisfying the needs of specific customers or markets. Very few companies, however, have the capacity, or even the desire, to satisfy all the diverse needs in their target markets. Thus, a set of criteria must be developed to screen for desirable business opportunities and some person or team must be assigned this screening responsibility.

Second, an R&D programme should be aimed at satisfying "generally known" needs in the target market(s). For example, all operators of air compressors want compressor oils that have longer operating lifetimes, although some of them are unwilling to pay the resulting higher prices. These needs may also be described as latent, not having been satisfied by any other source. Satisfying latent needs gives the firm the prestige of being first and a lead over competition in developing production volume and efficiency.

Third, an effective R&D programme should also attempt to expand upon those technologies that represent the firm's strengths. One of the factors leading to a product's potential success is the utilisation of technology that has already been proven in the manufacture of other products. Technologies can become obsolete, however, just as products do.

It is therefore necessary to evaluate which customer needs will be considered, to choose a technology capable of satisfying those needs and to recognise when the technology and needs are no longer compatible.

Industrial marketers must be especially sensitive to the changing needs of industry leaders, particularly if the industry is an oligopoly. When a few users or original equipment manufacturers (OEMs) represent a large portion of an industry's buying power, a supplier cannot afford to dissatisfy, let alone ignore, their needs. It should not be assumed, however, that current industry leaders will remain dominant. Thus, industrial marketers not only face the challenge of satisfying current industry leaders, but also must monitor and spot future industry leaders. This is a difficult task because future leaders often emerge in conjunction with, or because of, new technologies.

To support a strategic thrust, business objectives and marketing strategy must precede product development. The company must first decide what business it wants to be in and what quantitative goals it seeks to achieve. For example, a company may decide that each new product must be capable of sustaining a 15% compound growth rate over its life cycle while generating a 25% pretax return on assets. They should aim to either duplicate competitive products or make ones that stem from unique design or production techniques that would afford a differential cost and/or performance advantage. The growth criterion forces marketers to search out the most promising business opportunities, the demand for differential advantage minimises "me too" products while increasing the probability that products would be aimed at providing a measurable satisfaction level in specific applications and the profitability criterion serves as the cornerstone for financial analysis.

5.2.2 Idea Screening

Screening tries to eliminate those ideas that are likely to fail, tries to recognise those with promising potential and helps to optimise the remaining stages of the development process. It is important that the output of the screening stage (the number of ideas sent forward) be within the company's ability to act. Most businesses are not limited by the input of ideas but by the resources (time, money and/or staff) required to developed and successfully commercialise them.

5.2.3 Idea Evaluation

Even those ideas that pass the screening stage require further evaluation. If the idea originated internally, market need and potential will have to be determined. If the idea stems from a specific customer need, and the market criteria have been satisfied, it will have to be determined if the idea can be transformed into a physical product. When product ideas satisfy both market and product criteria, the firm should rank order the ideas. This ranking will vary with marketing strategy and business conditions and can be based on any combination of market, product and financial criteria. For example, the primary emphasis could be strengthening of market position, increased sales volume, improved profits, business diversification or a broader product range.

5.2.4 Agreement Between Marketing and Product Development Departments

Many companies, particularly smaller ones, have no formal process for evaluating product ideas. Additionally, they do not research the market to determine commercial feasibility before starting product design work. Given the increasing cost of product development and the increasing rate of technical and product change, both of which make product failures more costly, this intuitive approach could be regarded as unnecessarily risky. A potentially more profitable approach involves a written agreement between the marketing and R&D departments regarding the eventual product, in addition to the prerequisite research and evaluation. This agreement can be called a tentative data sheet, performance goals or product description.

Marketing should define these in terms of customer benefits, or what the product's attributes are expected to do for the user. These should be classified as essential, desirable and trade-offs. Essential benefits do not allow for compromise, representing the critical advantage(s) of the product. Desirable benefits remain so only if they do not detract from the essentials. Trade-offs are benefits that tend to impact each other. Price and total performance level are common trade-offs, and these may very well have to be renegotiated before the development process is completed.

Marketing inputs must be based on specific knowledge of customer needs, competitors' capabilities and general market conditions. As a result, marketing staff should limit their description of the physical product to those items essential for customer satisfaction or competitive positioning. Product development staff in the R&D department should be left free to utilise technology in whatever fashion will optimise performance and profitability. The agreement should help to protect R&D staff from constant changes in product performance goals or specifications.

5.2.5 Preliminary Business Analysis

After the above work, the company should have acquired sufficient information about customers, competitors, volume potential, tentative pricing, technology, investment level and estimated production cost to make a first-pass financial analysis. This analysis should be used to determine whether the initial idea should become a product.

The conversion of an idea into a product can be a significant portion of the total development cost, particularly for companies that require an elaborate pilot facility to prove production feasibility. In these instances, the preliminary business analysis is of major consequence. Other companies, whose development costs are relatively insignificant, may choose to skip this stage entirely on the basis that most of the numbers come from conjecture, forecast and guesswork. Some research, however, indicates that managers consider financial criteria the most important screening factors and are not likely to ignore them at this point.

Companies can face limits on resources, forcing them to prioritise and drop some otherwise promising projects. Combining the business analysis with other selection criteria provides the company with a means of choosing the best projects for further development, while the other projects are either dropped or put on indefinite hold.

5.2.6 Product Development and Testing

During this stage, R&D will convert the product idea into a physical reality, proving technical feasibility, while manufacturing will confirm or negate its ability to reproduce the product within the cost estimates and performance guidelines previously established. When product samples are available, marketing will approach selected customers to verify that the product's attributes do indeed satisfy specific application requirements. It is also important to reaffirm the market potential that was estimated earlier. During an extended development process, such as one that consumes a year or more, significant changes can occur in market and economic conditions, competitive capabilities and customer priorities. A product idea that was very promising a year ago may be virtually obsolete today.

Given the pragmatic and profit-oriented nature of industrial buying decisions, price plays a major role. In industrial markets, price therefore should be an important design criterion, in addition to product performance, quality and other factors discussed previously. Many companies, however, allow price to be a random effect rather than a specific goal. In these companies, costs are allowed to drive selling price rather than a target selling price in the market dictating a maximum acceptable cost. In many markets, both industrial and retail, target selling prices are dictated by competitors' selling prices for similar products.

5.2.7 Test Marketing

This phase involves the evaluation of the product by major potential users. Sales staff should have already identified these companies and all the staff who will evaluate the product and make or influence the buying decision. Acceptance of a new product by major users is neither automatic nor a rapid process regardless of the product's merits. A new component, for example, may require redesign of the end equipment to make full use of the cost or performance advantage. The equipment manufacturer (OEM) may not have personnel immediately available to do the redesign work. The production manager may convince other decision makers that it would be unwise to disrupt the smooth-running production line, since potential cost savings could be offset by lower product quality or reduced yields (due to the impact of change on production workers). Even if the decision is reached to make the necessary changes on a limited quantity basis, final approval may rest on the results of field tests. In other words, test marketing of the product could depend on test marketing potential buyers' products.

Such delays might be interpreted as a gloomy prospect for a new industrial product. On the contrary, if a product addresses an important need and satisfies it well, market acceptance will probably follow. Because time delays involved with test marketing can easily stretch to months, a new product supplier may face a period during which productive capacity is in place but no sizeable demand develops. Since some customers will have purchased limited quantities for evaluation purposes, the subsequent delay gives the impression of a market failure: a very brief growth followed by decline. Actually, the growth phase of the product life cycle has not yet begun.

5.3 FORMULATING AND DEVELOPING A NEW AUTOMOTIVE ENGINE OIL

Automotive engine oils are the most frequently reformulated types of oil and among the most complex, and so serve as a good example for the principles which can be applied in general to most types of oil. The steps involved in formulating and developing a new or improved lubricant are illustrated in Figure 5.2. These steps follow the same basic pattern as for any new or improved product, illustrated in Section 5.2.

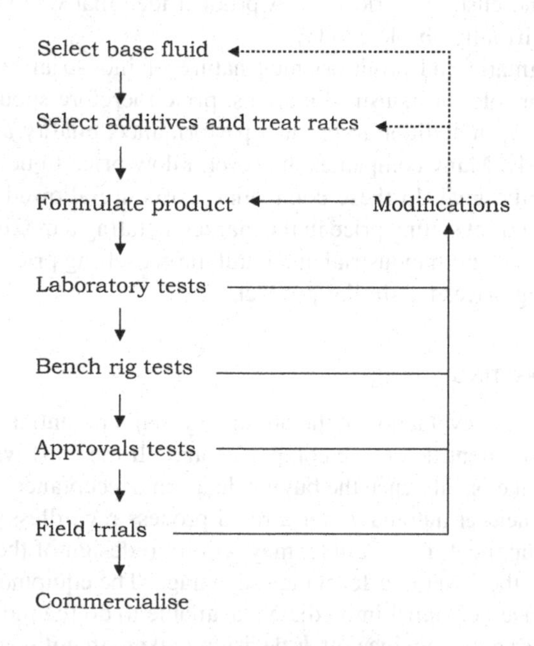

FIGURE 5.2 Developing a new or improved lubricant.

5.3.1 THE SPECIFICATION

This is the starting point of the formulation and development process and covers both the physical and chemical properties and the performance requirements, particularly in terms of test-passing criteria. For an engine oil, the performance may also be expressed loosely in general terms such as "turbocharged passenger car diesel engine oil". There may be other physical or chemical limitations, such as maximum zinc or phosphorus levels, and there may be appearance requirements, such as clear and transparent or dyed red.

International specifications for lubricants are published by the American Petroleum Institute (API), the Association des Constructeurs Européens d'Automobiles (ACEA), the International Lubricant Standardization and Approval Committee (ILSAC), the Japanese Automotive Standards Organisation (JASO) and the International Organization for Standardization (ISO). National specifications for lubricants are published by many organisations, including Deutsche Institut für Normung (DIN), Association Française de Normalisation (AFNOR), the American National Standards Institute (ANSI) and the American Gear Manufacturers Association (AGMA).

A large number of OEMs publish specifications for lubricants, including Ford, General Motors, BMW, Mercedes-Benz, Volkswagen, PSA, Renault, MAN, Volvo, Cummins, Caterpillar, Siemens, Parker, Bosch Rexroth, Atlas Copco, Ingersol Rand, General Electric, Rolls Royce and Mitsubishi.

Most specifications and oil formulations evolve out of earlier versions, and some ideas on a composition for the new oil usually exist based on prior technology. In the case of multigrade engine oils, the viscosity improver may have a greater effect on the engine cleanliness than the detergent or "performance" additives have on the viscometrics, and therefore the physical properties of viscosity at high and low temperatures are normally considered first. A "dummy" or best-guess performance additive treatment is used for this initial work.

Drivers for new lubricants, whether automotive, industrial, agricultural, mining, marine or other applications, include:

- OEM specifications.
- Industry specifications.
- Environmental issues.
- Health and safety issues.
- Improved energy efficiency demands.
- Revised or new applications.

Recent examples of these drivers include electric vehicles, extended oil lifetimes, biodegradability for oils used in outdoor applications and elimination of potentially harmful components or additives.

5.3.2 CHOICE OF BASE OIL(S)

Conventional solvent-refined paraffinic, API Group I base oils have been predominant in automotive engine oils for many years. However, with increasing performance requirements from engine and vehicle manufacturers, synthetic and hydroprocessed base oils, API Groups II, III, IV and V, are being used increasingly and are essential components for meeting many current specifications. An increasing number of lubricant marketing companies have found the ability to describe an engine oil as containing synthetic base oils a valuable aid to marketing and sales.

The "synthetic" base oils used in automotive engine oils are API Group III hydroprocessed mineral oils, polyalphaolefins and esters (both diesters and polyol esters). The reasons for including some or all of these in a formulation may range from a desire to promote a quality image and justify a higher price, to an inability to meet specification requirements without their use. Currently, such requirements include the restrictive volatility limits for SAE 0W, 5W and 10W multigrade oils contained in such specifications as those of the ACEA, API and ILSAC.

Which base oil(s) to use as the basis for an automotive engine oil formulation needs to be decided first. The same applies to automotive gear oils, automatic transmission fluids, industrial gear oils, hydraulic oils, turbine oils, air compressor oils and refrigerator oils. In some cases, for example, aviation gas turbine oils and natural gas compressor oils, there is little or no choice for base oils. For example, all current aviation gas turbine oils are based on polyol esters, most usually having a kinematic viscosity of 5 cSt at 100°C.

5.3.3 Choice of Viscosity Index Improver

For an automotive engine oil, the most important criteria in the selection of a viscosity index (VI)-improving additive are the level of shear stability and the level of diesel engine oil performance required. These are in conflict. Diesel engine oil performance is generally degraded by high polymer content, but improved shear stability is achieved by reducing the molecular weight of the polymer while using a higher treat rate (dosage) of polymer to provide the required viscosity improvement. The thermal stability of polymers is also important. Polyester VI improvers may be less suitable for severe diesel engine oil performance. For the most severe performance requirements, long-chain polymers with minimum side groups are used. They should be of the highest molecular weight, which will permit the shear stability targets to be met. Ethylene propylene and styrene isoprene copolymers are used most commonly in modern formulations, although new polymer types are being developed continually.

Polymethacrylate VI improvers tend to be used in industrial lubricants, particularly multigrade hydraulic oils and industrial gear oils. It is important that the VI improver does not cause excessive low-temperature thickening, which is likely to cause problems in the mini-rotary viscometer (MRV) or cold cranking simulator (CCS) tests, described in Chapter 10.

To formulate an automotive engine oil that meets the physical requirements, the viscosity and volatility of the various base oils, the thickening ability and shear stability of the VI improver and the influence of the pour depressant need to be determined. From this information, trial blends can be constructed using a dummy detergent additive package and submitted for tests against the specification. A typical, but not mandatory, order of testing would be

Property	If Failing, Modify
Kinematic viscosity at 100°C	Base oils, VI improver treat rate
Pour point	Pour point depressant
CCS dynamic viscosity	Base oils, VI improver type
MRV dynamic viscosity	Base oils, VI improver type, pour point depressant
Volatility	Base oils
Shear stability	VI improver, molecular weight
High-temperature high-shear viscosity	VI improver type and/or molecular weight

Descriptions and explanations of the various tests listed above can be found in Chapter 10.

5.3.4 Developing the Dispersant Inhibitor Package

Modifications to existing formulations of automotive engine oils may be required if there have been significant changes in the base oils, the viscosity improver or the performance specification requirements. The detergent inhibitors (DI packages) developed by each of the four major additive manufacturers differ slightly, but all contain a mixture of dispersants, detergents, anti-wear additives and oxidation inhibitors.

The type and treat rate of dispersants is normally set by the specification that has to be met for sludge performance. A number of engine tests are used to test for sludging in engines. A dispersancy credit may be available from the use of multifunctional dispersant viscosity modifiers to permit a reduction in the overall additive treat rate. Dispersants of good thermal stability are required for diesel performance, but the concentrations used are normally set by the sludge requirement. The thermal stability of a dispersant depends on the chemical type, its purity and the manufacturing method.

Originally, basic metal sulphonates were the principal soot-suspending components (detergents) in diesel engine oil formulations. However, with the development of highly overbased sulphonates and the incorporation of dispersants into diesel oils they are now seen primarily as a source of alkalinity. Base number (BN) can now be 400 or higher, while at the same time the soap content has tended to decrease. Magnesium sulphonates are sometimes now preferred to calcium, because they have lower ash levels for a given BN and the steel corrosion performance is superior. Sodium sulphonates have even better rust performance and lower ash, but can require extra anti-wear treatment due to their extreme affinity for metal surfaces. High treat rates of sodium-based additives have also been known to cause corrosion of aluminium pistons once these are de-oiled for servicing or test rating. The sulphonate molecule has a tendency to be pro-oxidation, so formulations containing high levels of sulphonate detergent require extra amounts of oxidation inhibitors.

Phenate and salicylate detergents have been used increasingly in automotive engine oil formulations. (Salicylates can be regarded as analogous to phenates, although with their extra carboxylic acid group they have two valencies available for bonding to metals, usually calcium.) Both these additives are powerful inhibitors and contribute a great deal to deposit control in diesel engines, especially in the upper piston areas. They also assist in preventing oil oxidation and thickening, particularly in their sulphurised forms. Phenates are available at several levels of BN, from unneutralised alkyl phenols (which are mildly acidic and react with some of the BN present from other additives) to 250 BN versions. The most popular additive is a 250 BN sulphurised calcium phenate.

In the past, zinc dialkyldithiophosphates (ZDDPs) compounds were used to provide the principal anti-wear and antioxidant properties of engine oil formulations. The chemical structure affects the anti-wear potency and the chemical stability of an individual ZDDP, and with four alkyl groups to every zinc atom many variations of structure are possible. In general, lower-molecular-weight and secondary alkyl types are less stable but have higher anti-wear activity, while high-molecular-weight primary types are more thermally stable but as a consequence have delayed or reduced anti-wear action. The former would tend to be used for speciality gasoline engine oils and the latter for diesel engine oils.

The older aryl (phenol-based) ZDDPs, which for a time were used extensively in diesel engine oils, are not now generally used, having been replaced by stable longer-chain primary types made, for example, from C_7 to C_9 alcohols. ZDDPs contribute to the ash level of a formulation, but a greater concern is their phosphorus contribution. Phosphorus is considered to be an exhaust emission catalyst deactivator, and many specifications now include maximum phosphorus limits.

The current limits for viscosity increase in some engine tests, coupled with phosphorus and ash limitations, mean that with most conventional base oils an oil antioxidant additional to any ZDDP is needed. There are many supplemental antioxidants commercially available, with the traditional oil antioxidants such as hindered phenols and aromatic amines most commonly used. Mixtures of two or more different antioxidant chemistries are often particularly effective because different oxidation mechanisms can be inhibited by each type. Many current automotive engine oil formulations now contain a combination of ZDDP, hindered phenol and amine antioxidants.

5.3.5 Evaluating and Finalising the Formulation

Before a new engine oil formulation can be considered ready to be marketed, it must first be tested against the required specification in a number of bench and engine tests. The very high cost of engine tests requires that care be taken in the order in which tests are run, so that late failures do not require rerunning too many tests with a revised formulation. Recent changes to pass/fail criteria ("statistical testing") and rules for the "reading-across" of prior results after minor changes have made the design of a test programme both more important and more difficult. An explanation of the way in which these tests are conducted and evaluated is beyond the scope of this book, but comprehensive information can be found on the API and ACEA websites and from each of the four main manufacturers of engine oil packages.

Normally the first properties to be considered are the physical properties, controlled by base oil selection, the use of synthetic or other special base oils, and the VI improver and pour point depressant additives. At this stage, a "performance package" or DI package is chosen for the commencement of testing. The knowledge of the formulator with regard to the performance of existing packages and the individual and combined responses of the components available when used in the various testing environments is crucial here. The formulator ideally should have response curves obtained from statistical testing for the key additive components. The availability of this information and the formulator's ability to use it both to decide on the initial package and then to modify it as necessary will very much determine the success and cost-effectiveness of the formulation programme.

When the physical properties have been adequately met, testing proceeds to the performance targets which are normally associated with standard engine tests. For some tests, there may be low-cost screening tests available, run either in simple laboratory equipment or in engines which are not qualified for approval testing. Tests are best run in order of increasing cost, so that if a test fails and the formulation has to be changed, it is the cheaper tests which have to be repeated. However, if new technology is being developed or a new quality level is being formulated, then those tests expected to cause problems would be run first in order to avoid late failures in a testing programme. In such cases, test order becomes a matter of personal judgement.

If a formulation change has to be made, it may not always be necessary to rerun all the earlier engine tests. For example, if there was a late failure in a major diesel engine test, then depending on the specific changes made, a certifying authority may allow "read-across" to a new formulation where the changes consist of small additions of already-present components or an increase in the total additive package. For example, it might be argued that a small dispersant addition would be generally neutral with oxidation and rust tests unaffected, while sludge control would be improved. On the other hand, an addition of extra sulphonate would improve the rust and the sludge performance but could possibly adversely affect the oxidation performance, and such tests would have to be rerun. In the case of a phenate addition, the oxidation performance would be improved, the sludge probably unaffected, but the rust performance might be harmed. A general increase in the amount of the total additive package used is normally considered beneficial, but care must be taken that restrictions such as maximum ash or phosphorus contents are not exceeded.

As soon as the broad structure of the new formulation is known, a representative blend can be made and subjected to tests such as corrosion resistance and anti-foam performance. Problems in these areas can often be fixed by the addition of compatibility agents, but it must be determined that a formulation can be blended before starting on expensive engine tests. The task facing an additive supplier, who must preferably incorporate the additives into a single concentrated package, is considerably more difficult than that of a lubricant manufacturer who may be prepared to blend the oil from individual additive components. Most engine oil formulations are, however, blended from whole or partial packages provided by additive manufacturers.

The foregoing descriptions on formulation may suggest that a formulator needs no more than a recipe book, some response curves and a few designed experiments to carry out these tasks. If so, the challenge has been understated. The interactions between components are extremely complex. Although some generalisation (such as effect of temperature or acid attack) will be equally relevant for different engine environments, nearly every new engine test brings unexpected challenges and is seldom introduced without lubricant reformulation.

Usually, it is not too difficult to meet the requirements of a single engine test in isolation. It is the combination of requirements that causes most problems. Conflicts and compromises can be caused by sludge performance versus corrosion inhibition or anti-wear performance, U.S. versus European or Japanese specifications or dispersancy and detergency versus anti-foam performance.

At the time of writing, the European ACEA specifications for light-duty gasoline and diesel engine oils included nine different engine tests, and for heavy-duty diesel engine oils there were seven different engine tests. With the U.S. API and ILSAC specifications, for light-duty gasoline and diesel engine oils there were six different engine tests and for heavy-duty diesel engine oils there were seven different engine tests.

A skilled formulator will know the ingredients and interactions but still occasionally fail to meet technical targets, irrespective of any cost constraint. The successes, like those of a master chef, sometimes seem to owe a little to art as well as a lot to science.

Once all the engine and performance tests have been passed and the required specification targets met, many automotive OEMs will require that a new engine oil be evaluated in one or more field trials. For example, Mercedes-Benz currently requires evaluation of a new engine oil in two different field trials, one of which takes 2 years and the other 3 years.

5.4 FORMULATING AND DEVELOPING A NEW INDUSTRIAL LUBRICANT

The general methodology for developing a new or improved industrial lubricant is essentially the same as that for an automotive engine oil, with some differences. It is still expensive and time-consuming.

Choosing the most suitable base oil(s) and additives is usually relatively easy. Once an initial "best-guess" formulation has been selected, the first step is to test it in simple, low-cost laboratory tests. A wide number of tests are performed to assess the physical or chemical properties of lubricants. Tests for physical properties include:

- Kinematic viscosity (capillary viscometer, low shear).
- Low-temperature viscosity: Brookfield viscometer (pumpability), CCS (crankability), MRV (pumpability).
- High-temperature viscosity: Tapered bearing simulator (high temperature, high shear [HTHS]), Ravenfield viscometer.
- Pour point.

- Flash point: Pensky–Martens closed cup, Cleveland open cup.
- Volatility: NOACK, gas chromatography (GC) (simulated distillation), air jet, distillation.
- Foaming tendency and stability.
- Density (specific gravity).

Tests for chemical properties include:

- Acidity or alkalinity: Acid number, neutralisation number, BN.
- Sulphated ash.
- Elemental analysis: Flame photometry, atomic absorption (AA) spectroscopy, inductively coupled plasma (ICP) emission spectroscopy, x-ray fluorescence (XRF) spectroscopy.
- Infrared (IR) spectroscopy.

Only some of the following tests will be of practical value in a lubricant blending plant. They are mainly used to develop new or improved lubricants.

It is particularly important to emphasise here that the only way in which to determine whether a lubricant will satisfactorily lubricate the machinery or equipment for which it was designed is to use it in that machinery or equipment for a number of years. The purpose of the development tests is to identify those candidate formulations that are unlikely to work in practice. The cheapest and quickest tests are done first. Candidate formulations that pass these tests are then subjected to longer, more demanding (and inevitably, more expensive) tests.

After laboratory tests come bench tests, such as:

- Oxidation resistance: IP 280, IP 306, ASTM D943, ASTM D2893, PDSC (ASTM D6186), TEOST (ASTM D6335).
- Thermal stability: Panel coker (FTM 3462).
- Shear stability: Diesel injector (IP 294), tapered roller bearing.
- Corrosion resistance: Steel (IP 135, ASTM D665), copper (IP 154, ASTM D130), high-temperature corrosion bench test (HTCBT) (ASTM D5968).
- Anti-wear, load-carrying or extreme-pressure properties: Timken (ASTM D2782), Falex (ASTM D2670), four-ball (ASTM D2783), pin-on-disc, FZG (IP 351).

Then come machinery or equipment tests. For industrial lubricants, there are numerous machinery and equipment tests required to meet OEM specifications.

For hydraulic oils, the most commonly used tests include those in vane pumps, piston pumps, the FZG gear rig and a simulated hydraulic system. Industrial gear oils will need to pass the highest load stage in an FZG gear test. Most industrial lubricant formulation laboratories will like to test a new or improved air compressor oil in either a stationary rotary screw compressor or a stationary piston compressor rig. Steam turbine oils are not usually tested in laboratory equipment but pass straight to evaluation in a "real" turbine system. The same applies to refrigerator compressor oils, reactive gas compressor oils and transformer oils.

At the conclusion of these tests, many OEMs require that a new oil undergo one or more field trials, which can last between 1 and 3 years.

The purpose of Figure 5.2 is to show that if one or more tests fail at any stage of the process, the only solution is to make modifications and then start the process again from the product formulation stage. If the test failure is particularly significant, the process may need to be started afresh from either the base oil selection or additive selection stages. Too many corrections or modifications add to time and cost.

It is not uncommon for the entire process to take 2 or 3 years from beginning to end. In some cases, particularly for new automotive engine oils, the process can take 5 or 6 years.

Blending plant managers and supervisors need to be aware that developing and testing a new lubricant formulation is a complex, demanding, time-consuming and expensive exercise. Even small errors in blending the precise amounts of components in any formulation can have significant consequences for the eventual performance of the lubricant in practice.

5.5 ILLUSTRATIVE LUBRICANT FORMULATIONS

To demonstrate the differing levels of complexity of oils that might be manufactured in a blending plant, we have included a number of illustrative formulations for various types of lubricants. These are shown in Tables 5.1 through 5.10.

The purpose of these illustrative formulations is to show the range of base oils and additives that are used in different types of lubricants; that some of the formulations are quite simple, while others are quite complex; and that some of the lubricants do not contain any base oil(s) at all. It should be noted that some of the formulations are relatively dated. For example, the sodium borate in water-mix metalworking fluids (Table 5.7) is being increasingly replaced by less environmentally damaging compounds, as is the use of chlorinated hydrocarbons in neat cutting oils (Table 5.9).

It should be noted that these formulations are illustrative only and are not intended to be copied or used as substitutes for a carefully managed R&D programme. There is no reference as to which OEM or industry specifications these formulations may or may not meet. However, they might be useful as a starting point for such an R&D programme.

TABLE 5.1
Illustrative Formulation for a Gasoline and Light-Duty Diesel Engine Oil

Component	% wt
150 SN Group I base oil	77.5
DI package	12.0
VI improver	10.0
Pour point depressant	0.5

Source: Pathmaster Marketing Ltd.

TABLE 5.2
Illustrative Formulation for a Multigrade Automotive Gear Oil

Component	% wt
PAO 6	55.0
4 cSt Group III base oil	19.6
Extreme-pressure additive package	4.4
PIB VI improver	18.0
Ester	2.0
Pour point depressant	1.0

Source: Pathmaster Marketing Ltd.

TABLE 5.3
Illustrative Formulation for an Anti-Wear Hydraulic Oil

Component	% wt
150 SN Group I base oil	96.95
Oxidation inhibitor	1.0
Anti-wear additive	1.0
Corrosion inhibitor	0.5
Metal passivator	0.05
Pour point depressant	0.5

Source: Pathmaster Marketing Ltd.

TABLE 5.4
Illustrative Formulation for a Steam Turbine Oil

Component	% wt
150 SN Group I base oil	97.95
Oxidation inhibitor	1.5
Corrosion inhibitor	0.5
Metal passivator	0.05

Source: Pathmaster Marketing Ltd.

TABLE 5.5
Illustrative Formulation for an Air Compressor Oil

Component	% wt
500 SN Group I base oil	97.95
Oxidation inhibitor 1	1.0
Oxidation inhibitor 2	1.0
Corrosion inhibitor	0.5
Metal passivator	0.05

Source: Pathmaster Marketing Ltd.

TABLE 5.6
Illustrative Formulation for a General-Purpose Soluble Oil Emulsion Metalworking Fluid

Component	% wt
100 SN Group I base oil	68
Sulphonate emulsifier base	17
Chlorinated olefin	5
Synthetic ester	5
Alkanolamide	3
Biocide	2

Source: Pathmaster Marketing Ltd.

TABLE 5.7
Illustrative Formulation for a Biostable Microemulsion Cutting Fluid

Component	% wt
MVI 100 naphthenic oil	15.0
Nonyl phenol ethoxylate	12.0
Sodium naphthenate	10.0
Polymer ester	5.0
Ethylene glycol	50.
Butyl glycol	2.0
Polypropylene glycol	1.0
Laurylamine	8.0
Sodium borate	10.0
Benzotriazole	0.1
Water	31.9

Source: Pathmaster Marketing Ltd.

TABLE 5.8
Illustrative Formulation for a Synthetic Cutting Fluid

Component	% wt
Amine carboxylate	10.0
Triethanolamine	5.0
Polyethylene glycol ester	5.0
Phosphate ester	4.0
Sulphated castor oil	4,0
Pyridinthione	2.0
Blue dye	0.01
Water	69.99

Source: Pathmaster Marketing Ltd.

TABLE 5.9
Illustrative Formulation for a Neat Cutting Oil

Component	% wt
MVI 100 naphthenic oil	90
Lard oil	2
Chlorinated paraffin	6
Sulphurised lard oil	2

Source: Pathmaster Marketing Ltd.

TABLE 5.10
Illustrative Formulation for a Sulphurised Extreme-Pressure Gear Cutting Oil

Component	% wt
100SN Group I base oil	58
Sulphurised fatty ester	25
Rapeseed oil	15
Polysulphide	2

Source: Pathmaster Marketing Ltd.

5.6 EASE OF BLENDING

As illustrated in the previous section, some products developed by lubricant formulation chemists turn out to be relatively simple and easy to blend. They generally involve only one grade of base oil, or perhaps two base oils and two or three additives.

All components are available as liquids. In the most easily blended products, all base oils and additives have comparatively low viscosities. Example products include:

- Automotive and industrial engine oils.
- Monograde industrial gear oils.
- Hydraulic oils.
- Turbine oils.
- Compressor oils.
- General-purpose neat metalworking fluids.

Other products prove to be a great deal more complicated to blend. These usually involve a number of additives (one specific formulation developed by the author contained 15 additives), some of which have high viscosities or are only supplied as solids. These formulations involve having to prepare a number of preblends before final blending can be initiated. Example products that are often more difficult to blend include

- Multigrade automotive and industrial gear oils.
- Automatic transmission fluids.
- High-performance neat metalworking fluids.
- Water-mix metalworking fluids.
- Lubricant dispersions and pastes.
- Electrical and cable oils and compounds.

Because of these issues, formulation chemists need to be aware of the possible problems they may be creating for a blending plant. Questions a formulation chemist should consider include:

- Are the specific additives used in the formulation easy or difficult to blend?
- Are any of the additives oxidatively or thermally unstable?
- Are any of the additives likely to react with each other during blending?
- Could a liquid additive be used, rather than a solid additive, while not affecting the properties or performance of the finished lubricant?

Formulations that are easier to blend can make life much easier for managers, supervisors and operators of blending plants.

5.7 COMMUNICATION AND COOPERATION BETWEEN FORMULATORS AND BLENDERS

A very important issue arises from the information outlined in the preceding sections; lubricant formulators and blending plant supervisors need to communicate and cooperate actively and effectively. Lubricant formulators should know how blending plants and blending equipment function. They should work toward making a potential formulation as easy to blend as possible. Lubricant formulators should consult with lubricant blenders about potential problems.

At the same time, blenders of lubricant should accept that formulating a new or improved lubricant can be very complex and difficult, involving choices between base oils and between additives. Lubricant blenders should also know that minor changes to formulations can significantly influence finished lubricant properties and performance. As illustrated in Chapter 4, additives cannot be simply swapped in a specific formulation. If an additive specified in a formulation is not in stock, that lubricant cannot be manufactured until it is in stock.

Two areas of potential concern arise from the possibility of what appears to be minor changes to a formulation.

A dumbbell blend involves mixing a low-viscosity base oil with a high-viscosity base oil to achieve an intermediate-viscosity lubricant. A blending plant manager or supervisor may be tempted to mix a low-viscosity base oil with a high-viscosity one if the stock of a medium-viscosity base oil is running low. For some lubricants, this does not present any problems. For other lubricants, it is likely to cause very serious problems with the performance of the product. Examples include:

- Low-viscosity multigrade automotive engine oils.
- Compressor oils.
- Steam turbine oils.

The problems arise when the low-viscosity base oil, which is more volatile than the high-viscosity base oil (see Chapter 2), begins to volatilise from the engine, compressor or turbine. A higher proportion of the high-viscosity base oil will be left in the machine, causing the oils' viscosities to increase. This is likely to result in problems with increased oxidation, increased sludging tendency and lower energy efficiency.

Another issue is that lubricant formulators and blenders are sometimes tempted to add just a bit more additive, additives or additive package, just to make sure that a finished lubricant meets all the performance requirements and/or specifications.

This is called "product giveaway". It means that the formulation or blend is more expensive than it should be, thereby reducing the product's profit margin. Too much additive means less profit, but too little additive means a product may cause problems for customers. Accuracy in blending is therefore very important

5.8 SUMMARY

Lubricant formulations can be simple or complex and easy or difficult to blend. Formulating a lubricant involves many complicated choices, including making it as easy to blend as possible.

Lubricant formulators and blenders need to communicate and cooperate to understand each other's problems. Accuracy in blending a formulation is very important.

6 Lubricant Blending Plant Design
Grassroots Plants and Upgrading Existing Plants

6.1 INTRODUCTION

Most finished lubricants, except greases and some water-mix metalworking fluid concentrates, are produced by mixing one or more base oils with one or more additives, in a procedure known as "blending". The base oils can be mineral oils or synthetic functional fluids, as discussed in Chapters 2 through 4, or mixtures of compatible mineral and synthetic oils.

As illustrated in Chapter 5, the proportions of base oils and additives vary for different types of lubricants. Many industrial lubricants, such as hydraulic oils, compressor oils, turbine oils, gear oils and general machinery lubricants, contain between 0.5% and 5.0% of additives blended into between 95.0% and 99.5% of base oils. Automotive engine oils can contain between 5.0% and 25.0% of additives. Some synthetic water-mix metalworking fluid concentrates can be 100% "additives".

Clearly, the procedures for blending lubricants, together with their subsequent packaging and storage, are critically important to the delivery of products of the correct quality and performance to customers.

6.2 INVESTMENT IN BLENDING LUBRICANTS

6.2.1 Business Aims

Investment in a lubricant blending plant should be supported by a number of essential business aims:

- *Variety of products*: Most lubricant marketers will aim to sell a range of products to meet market demands. This means that their lubricant blending plants will be required to make these products. The number of different formulations can vary from less than 100 in developing markets to more than 1000 in developed markets. The products may be divided into between about 8 and 25 product families and may be delivered in many different package sizes, ranging from 0.5 L bottles to bulk tankers.

- *Improved productivity*: This is achieved using automated systems. Until recently, the use of automation was regarded as only a production aid that was the responsibility of the information technology (IT) department. Now, every company needs to understand the impacts and benefits of new ITs, so they can achieve significant and sustainable competitive advantages.
- *Diversity of raw materials*: Simple lubricant formulations may involve one base oil and one or two additives or additive packages. More complex formulations may involve three base oils and 10 additives, some in drums, others in bulk and others as powders. One blending plant may need to accommodate 10–15 different grades of base oils and several hundred different additives and additive packages.
- *Use of additives*: The use of additives has a significant impact not only on the quality and performance of the product, but also on its costs. It is therefore essential to deal economically with these raw materials. As explained previously, overdosing needs to be avoided.
- *Contamination*: Blending plants are required to produce many different lubricants, but contamination between the various products must be avoided at all costs. In the past, separate tanks and pipelines were required or intensive flushing between the blends had to be done. This necessity limited the flexibility of the blending plant and considerably increased the running costs through the generation of slop oil.

As will be explained and discussed in more detail in Chapters 11 and 14, a blending plant is part of a lubricant supplier's marketing, sales and supply chain strategy and plans. It therefore needs to be completely integrated into this strategy.

6.2.2 Blending Plant Conception

The theory of blending plant operation as part of supply chain management needs to be converted into practical reality. The main difficulty for designers and constructors of blending plants is to translate the body of conceptual knowledge into a system and set of processes that may not necessarily be ideal for everything that may be required of the blending plant in the future.

When dealing with such a complex task, it is essential to define the precise information that will need to be used by the designer(s) of a blending plant so that they can prepare a realistic conceptual design. This phase is very important, since planning mistakes can lead to considerable additional costs if changes are required later.

For this purpose, numerous items of data need to be collected and evaluated:

- The mix of finished products.
- The types of packages to be used for each grade of product.
- Stock policy for finished products; readiness for dispatch.
- Numbers, types and amounts of raw materials.
- Types of and schedules for delivery of raw materials; bulk, drums and sacks.
- Stock policy for raw materials.

- Families and types of product formulations, together with typical and special blending procedures.
- Set-up times for each type of formulation or blend.
- Blending time for each type of formulation.
- Types of packages and filling methodologies for finished products.
- Methods and schedules for dispatch of finished products.
- Numbers and types of employees in the blending plant.
- Costs of blending equipment.

Based on these data, the optimum batch size (OBS) for each product type to be produced in the blending plant can be calculated using a mathematical formula:

$$OBS = \sqrt{\frac{2 \times Q \times OC}{SC}}$$

where:

Q = Annual production requirement, in tonnes
OC = Order cost, in dollars per blend
SC = Annual storage carrying cost for the product, in dollars per tonne

To illustrate how to use the OBS calculation, consider a requirement to manufacture 7000 tonnes per year of a specific automotive lubricant. (Illustrative numbers will only be used, since each blending plant will be different, depending on the market, location and country in which it is located.)

The order cost for one batch of this product is $100 per blend. The annual economic storage carrying cost of the product can be calculated from the storage cost per tonne (or kilogram) multiplied by the carrying cost, which is the assumed cost of capital. For the example blend in the specific blending plant, the unit storage cost is $6.21 per kilogram and the carrying cost is 7.5%, so the annual storage cost for this product is $(6.21 \times 1000) \times (7.5 \div 100) = \$465.75$ per tonne.

The OBS for the example is therefore the square root of $2 \times 7000 \times 100 \div 465.75 = 54.83$ tonnes. For this automotive lubricant, blends should probably be done in either 55- or 60-tonne batches.

Obviously, each product type in each product group manufactured by a blending plant will have a different OBS, since this will depend on the individual annual sales volumes for each product in the lubricant marketer's target market. A graphical illustration of the rationale behind the calculation of the OBS for each product is shown in Figure 6.1.

The formula can also be used to calculate the minimum required stock capacity for base oils and finished products and supports the decision for the choice of which blending equipment, discussed in Chapter 7, to be used.

Long before the terms just-in-time (JIT), total quality management (TQM), and manufacturing resource planning (MRP) were "invented", companies were using these concepts in managing their production and inventories. The basic concepts were published in 1931 in *Purchasing and Storing*, a textbook that was part of a modern business course at the Alexander Hamilton Institute in New York.[1] The

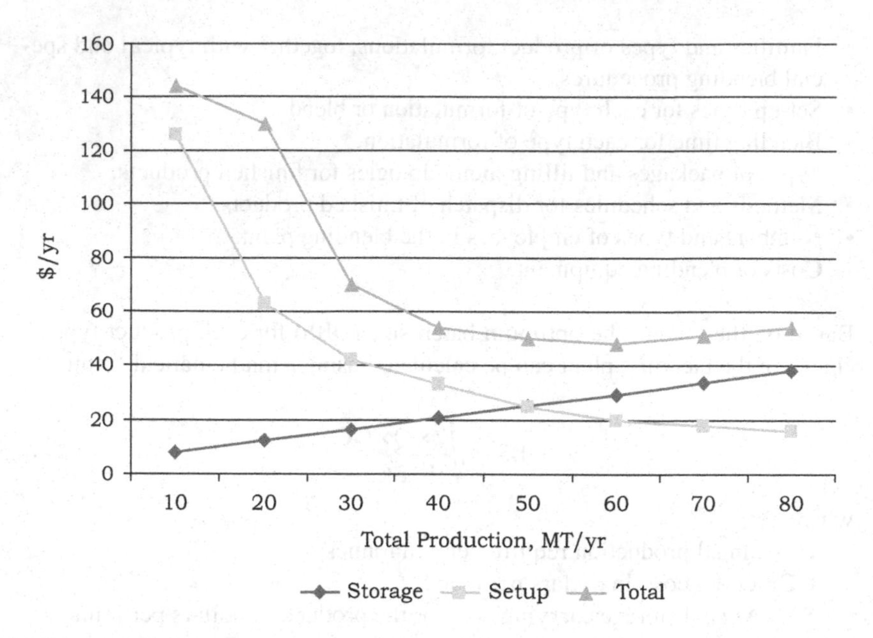

FIGURE 6.1 Illustration of optimum batch size determination.

textbook was essentially a how-to book on inventory management in a manufacturing environment. JIT, TQM and MRP are discussed in detail in Chapter 14.

It may appear that OBS conflicts with JIT. Although JIT is often described as arranging to have all components arrive in the exact run quantities just in time for the production run, it is actually a quality initiative with the goal of eliminating wasted steps, wasted labour and wasted cost. OBS should be one of the tools used to achieve this. OBS is used to determine which components fit into this JIT model and what level of JIT is economically advantageous for a specific operation.

To determine the most cost-effective quantities of batches for each type of product, use of the OBS formula will allow the determination of which types of equipment should be selected for a lubricant blending plant.

6.3 GRASSROOTS BLENDING PLANT

The input information required to design a grassroots (brand new) lubricant blending plant is slightly different than the information required for an upgrade to an existing plant.

Of most importance is, what is the lubricant manufacturer's marketing, sales and production strategy? Questions such as the following need to be analysed and answered:

- What will be the annual throughput of the new blending plant?
- What types of and how many products are required to meet market demands?

- From where and how are the raw materials going to be delivered; bulk by ship or barge, road tankers or rail tank wagons, drums and sacks?
- To where and how are the products going to be dispatched; direct to customers or to a distribution warehouse; by barge, road and/or rail?

Then the location and facilities need to be determined:

- Where is the blending plant going to be located?
- What space is available at this location?
- Does this location have road, rail, barge and/or ship access?
- What and how many blending units are going to be needed to meet the throughput and product targets?

All this information can then be used as input to the computer blending plant design programme.

6.4 UPGRADING AN EXISTING BLENDING PLANT

When considering whether and, if so, how to upgrade an existing blending plant, the questions that need to be addressed are somewhat different. First, what is the objective? This will need answers to the following questions:

- Will the plant need new or improved equipment?
- Is the upgrade simply to automate the existing units?
- Is the upgrade intended to generate a higher throughput?
- If so, will more equipment by needed?
- Is the upgrade intended to manufacture a wider range of products?

If a bigger blending plant is required, then further questions arise:

- Can the building footprint be enlarged?
- Can the buildings be higher?
- Will a bigger warehouse be needed?
- Is there enough space for more tanks?
- What will be the effects of more trucks, trains and/or barges?

The available space may limit the size of the upgrade or may prompt consideration about whether to build a new, larger blending plant in a different location.

6.5 BLENDING PLANT LAYOUT

A blending plant has a number of component parts:

- Base oil tanks.
- Additive tanks.

- Drummed additive storage.
- Pipe racks with piggable lines.
- Blending building.
- Empty package storage.
- Pack filling line.
- Bulk product storage tanks.
- Filled pack storage.
- Warehouse.

An illustration of the relationships between each of these items is shown in Figure 6.2.

The storage tanks obviously need to be outdoors. More detail about storage tanks for base oils, additives and finished lubricants is provided in Chapter 13, which will also discuss the management and operation of the warehouse. Details about the other items will be discussed in Chapter 7.

The storage areas for additives in drums or sacks, empty packages and filled packages of finished lubricants could be in the same building or in separate buildings. The advantage of separate buildings for storage is that there is no confusion about what is stored where. The obvious disadvantage is cost; one large building is likely to be considerably cheaper to both build and operate. Also, one large warehouse is likely to be much easier to automate.

The blending units and package filling lines could also be in a single building, in two buildings or in the same building as the storage areas. Again, cost to build and operate is a factor. However, having the blending units, where some additives could be used as powders, in the same building as the filling lines may risk cross-contamination.

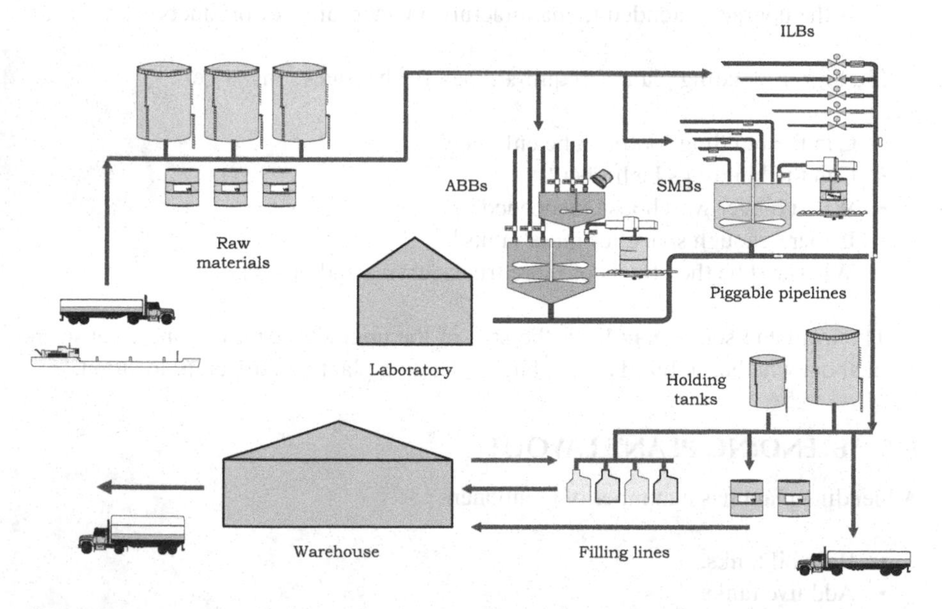

FIGURE 6.2 Relationships between components of a lubricant blending plant.

With regard to the blending units, these can be either an automatic batch blender (ABB), a simultaneous metering blender (SMB) or an in-line blender (ILB). The blending area will also benefit from having a drum decanting unit (DDU) and a drum heating unit (DHU). A large blending plant may require several ABBs, SMBs, ILBs and DDUs. All these items of equipment will be explained and discussed in Chapter 7.

These components can be arranged in one of two principal ways: in a linear manner or in a circular layout, as shown in Figures 6.3 and 6.4.

In the linear layout, the flow of materials through the blending plant proceeds from raw material delivery at one end of the plant to a raw materials storage area, comprising tanks and a drum storage warehouse. This storage warehouse may also include an area for storing empty drums and bottles. Raw materials are then fed into the blending and filling area, which may include intermediate holding tanks. From the blending vessels and package filling lines, products are passed to the finished

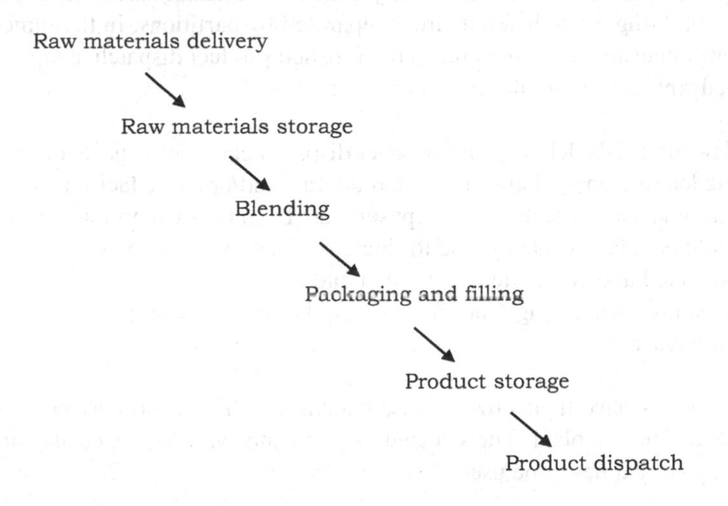

FIGURE 6.3 Flow of materials through a lubricant blending plant: linear operation.

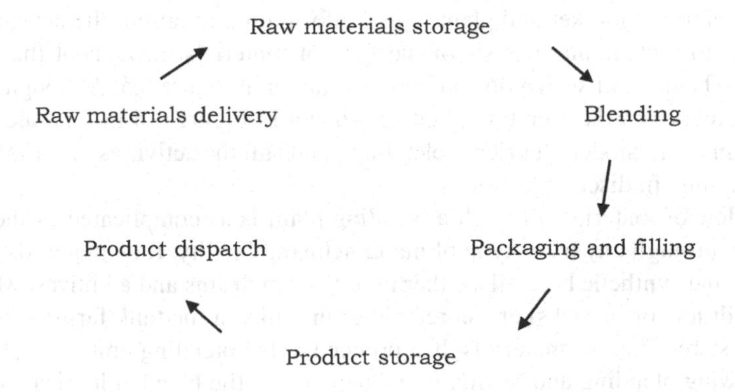

FIGURE 6.4 Flow of materials through a lubricant blending plant: circular operation.

product storage tanks and packaged product warehouse. At the other end of the plant from raw material input is the finished product dispatch area.

The advantages of linear layout are:

- Distances between operations, for operators, can be as short as possible.
- Pipeline transfers can be as short as possible; pigging operations are easier.
- There is no risk of cross-contamination or misidentification between raw materials and finished products.
- Raw material delivery and product dispatch can be kept separate.

In the circular layout, the flow of materials starts with raw material delivery and proceeds as with the linear layout, but the farthest point of the plant is the blending area. In between is a large warehouse for the storage of raw materials in drums and sacks, empty drums and bottles ready for filling and filled bottles and drums of finished products. The filling line is also adjacent to the blending area. All this can be in the same building, with different areas separated by partitions. In the same vicinity as the raw materials receiving area is the finished product dispatch area.

The advantages of circular layout are:

- Raw materials delivery and product dispatch can, with adjacent unloading and loading bays, share the same road, rail and/or barge facilities.
- Raw material storage can be kept separate from finished product storage but could be located side by side for increased flexibility. (Storage areas can be expanded and contracted to suit demand.)
- Blending, packaging and filling can be in a different building from warehousing.

Both layouts have their obvious disadvantages. There is no perfect or preferred layout for a blending plant. The selected location and available space may dictate the type of layout that has to be used.

6.6 DESIGNING A LUBRICANT BLENDING PLANT

In addition to the market and plant layout information required, the design process will need to include an analysis of the flow of materials throughout the blending plant. A schematic of such a flow analysis is shown in Figure 6.5. Although the individual activities in the blending plant are shown in Figure 6.5 in separate units, in practice in many modern lubricant blending plants all the activities occur in one very large building, in discrete locations.

The flow of materials through a blending plant is as complicated as the flow of materials through any other type of manufacturing facility. Raw materials (mineral base oils and synthetic base oils, either in bulk or in drums and additives, whether in bulk, in drums or in sacks) are stored either in tanks in the tank farm or in the raw material store. The raw materials then proceed to the blending units.

Following blending and testing (see Chapter 10), the blended lubricants are sent either to holding tanks in the tank farm or direct to filling lines. The holding tanks in

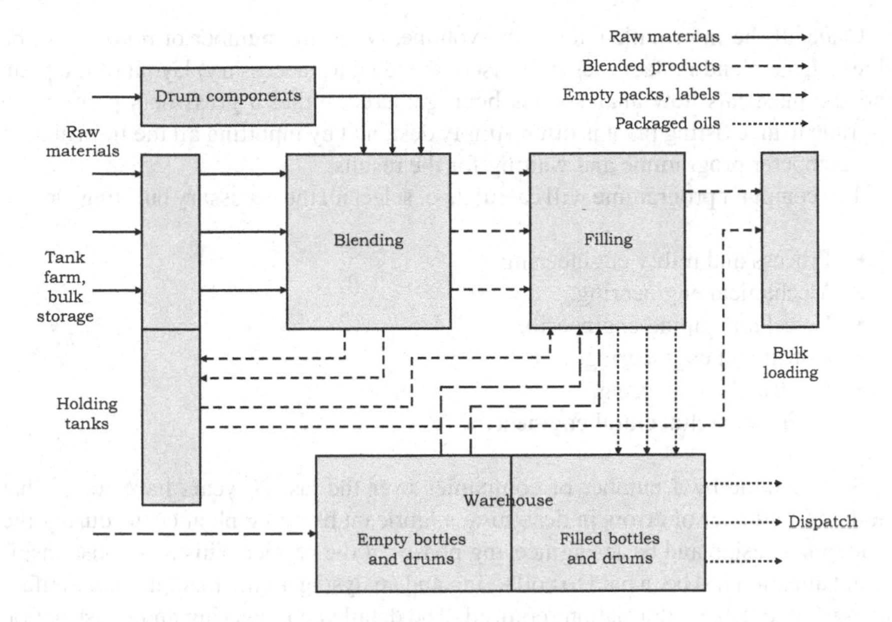

FIGURE 6.5 Lubricant blending plant materials flow analysis.

the tank farm are separate from those used to store raw materials in bulk. Products from the holding tanks are then sent to either the filling lines or bulk loading into road tankers, rail tank wagons or shipping vessels.

In the filling area, blended lubricants are filled into plastic bottles, drums or intermediate bulk containers (IBCs) or other containers (see Chapter 11). Empty plastic bottles, drums and IBCs are stored in the blending plant warehouse, ready for sending to the filling lines. In many modern lubricant blending plants, blow moulding facilities are used to manufacture plastic bottles on site, rather than having hundreds of thousands of empty plastic bottles (filled with air and weighing very little) delivered by truck. The plastic raw materials, empty drums and empty IBCs are delivered to the warehouse beforehand. Some lubricant blending plants have facilities to clean and reuse drums and IBCs (see Chapter 11).

Also in the empty packages section of the warehouse, room must be available for the labels that are going to be attached to plastic bottles following filling and capping, prior to the bottles being put into cardboard boxes. That means that the materials with which to make empty cardboard boxes must also be delivered to and stored in the warehouse.

At the other end of the warehouse, filled and packaged plastic bottles, drums, IBCs and other containers need to be stored, before delivery to customers, typically in trucks. Empty pallets for drums, IBCs and plastic bottles in cardboard boxes may also need to be stored in this part of the warehouse, or in the empty package section of the warehouse. Pallets may need to be used in other parts of the lubricant blending plant, for transporting base oils or additives in drums to the blending area or for transporting filled drums from the filling area to the storage area in the warehouse.

Once all the information about the volume, types and number of products to be blended, the blending facilities to be used, the ideal (or necessary) layout of the plant and the materials flow analysis has been gathered, either a grassroots plant or an upgrade to an existing plant is often simply designed by inputting all the information to a computer programme and waiting for the results.

The computer programme will calculate or select all the necessary building blocks:

- Process and utility engineering.
- Mechanical engineering.
- Vessel and piping engineering.
- Instrument engineering.
- Electrical engineering.
- Civil and architectural engineering.

Studies done by a number of companies over the last 20 years have found that the largest number of errors in designing a lubricant blending plant occur during the conceptual design and basic engineering phases of the project. This is because insufficient attention has been paid to collecting and analysing all the marketing, manufacturing and logistics information required. The detailed engineering and construction phases of the project produce a smaller number of errors in designing the plant.

Unfortunately, most of these errors do not become apparent until the managers, supervisors and operators in the blending plant try to make it work. Correcting design errors during blending plant operation has been found to cost more than twice the cost of correcting design errors found during construction or in detailed engineering.

When all the building blocks have been fine-tuned to ensure that they can be integrated and will function as intended, the computer programme will produce all the detailed drawings and instructions automatically.

The relationships between the various design elements for a lubricant blending plant are shown in Figure 6.6.

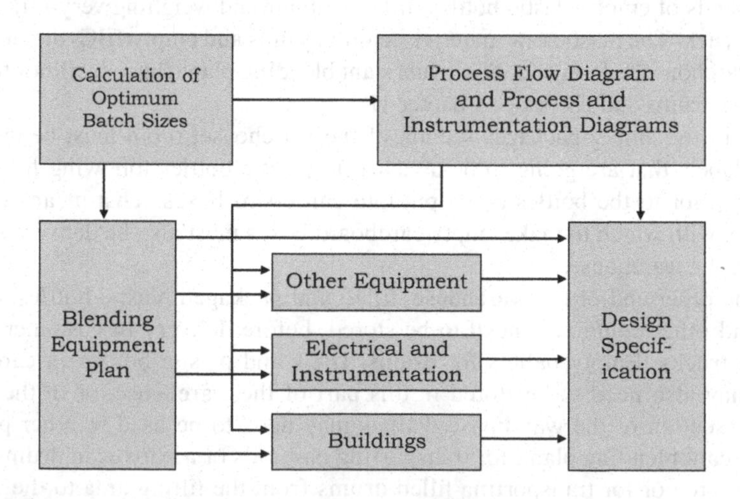

FIGURE 6.6 Design elements of a lubricant blending plant.

Specialist companies have developed these sophisticated computer programmes to the extent that a client lubricant marketing company can view a three-dimensional (3D) visual representation of what the blending plant will look like. They can also take a computer-generated tour through the plant's various buildings and areas. Two 3D computer-generated conceptual layouts for a lubricant blending plant are shown in Figures 6.7 and 6.8. The layouts are for the same plant, but from different perspectives.

Companies that produce the computer-generated design for the blending plant are likely to also provide full engineering, procurement and construction (EPC) services as well. They will work with the client lubricant manufacturer as part of the team

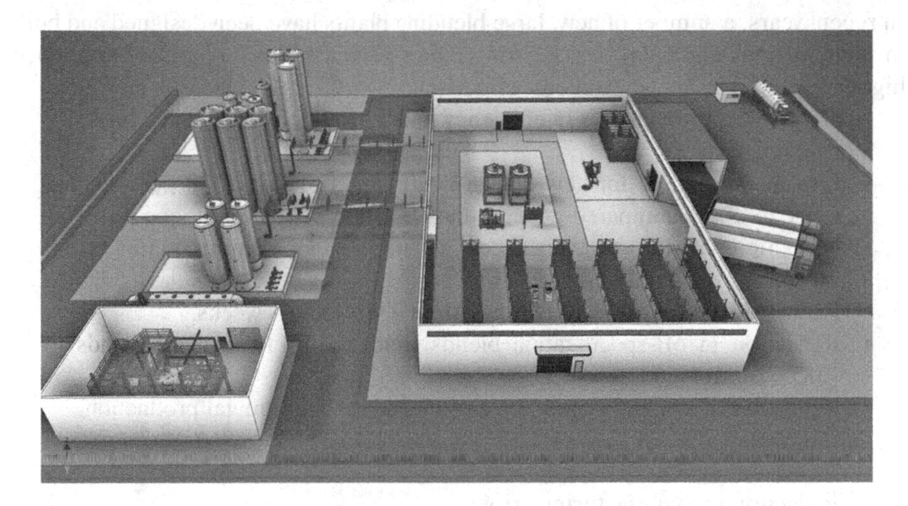

FIGURE 6.7 3D computer-generated conceptual layout for a lubricant blending plant. (From Fluid Solutions GmbH. With permission.)

FIGURE 6.8 3D computer-generated conceptual layout for a lubricant blending plant. (From Fluid Solutions GmbH. With permission.)

involved with any or all of the critical tasks, such as process design, plant layout and piping, skid-mounted packaged processing units, blending vessels, instrumentation and control systems and facilities and process management software.

Many of the design engineering companies will have partner companies in many countries or regions worldwide. Although many of the design engineering companies are headquartered in Europe or the United States, they will be able to provide services in Asia, South America, the Middle East and Africa.

6.7 BENEFITS OF HIGH-QUALITY BLENDING PLANT DESIGN

In recent years, a number of new, large blending plants have been designed and built in many regions and countries. Existing plants have also been upgraded to a much higher level of consistency and cost-effectiveness.

Examples of improvements in efficiency and cost-effectiveness include:

- Capital investment for new (green field) plants of <US$300 per year per tonne in 2006, compared with around US$500 per year per tonne in 1997.
- Reduction of the time between planning and commissioning from 24 months to 14 months.
- Reduction in commissioning times from 7 weeks to 4 weeks.
- Reduction in off-specification blends from 5% to 10% of all blends to <0.5% of all blends.
- Reduction of slop-oil generation from 3% to <1% of annual production.
- Improved output.
- Increased regulatory compliance.
- Reduction in manufacturing errors.
- Real-time operational decision support.

There is no doubt that modern computer programmes and blending plant automation have contributed extensively to these benefits.

6.8 SUMMARY

Designing a grassroots blending plant or upgrading an existing one involves more than just process and design engineering.

Flexibility is defined in the conception phase and determined in high-quality design and process engineering. With the use of computer-aided planning methods and standard tools, a cost-effective design can be achieved for new investments and upgrades.

Because a blending plant is one part of a complete supply chain, the facilities and operation of a blending plant must be part of an integrated marketing, sales, production and distribution strategy.

Once a large amount of marketing and supply chain information has been gathered and analysed, computer programmes can be used to generate the detailed design of the blending plant. All this makes it possible for a participant in the lubricants market to invest quickly in new products and markets and develop a higher market share in existing markets.

REFERENCE

1. *Purchasing and Storing: Modern Business Course*, Alexander Hamilton Institute, New York, 1931. (Not in print; the Alexander Hamilton Institute was dissolved in the 1980s.)

7 Lubricant Blending Plant Equipment and Facilities and Their Operation

7.1 INTRODUCTION

Many people in the lubricants business regard lubricant blending as simply mixing a few raw materials together to produce finished lubricants. The processes involved are not particularly complex or difficult to control.

The process of producing finished lubricants from base oils and additives is invariably described as oil blending rather than oil manufacture because there is no significant chemical reaction which takes place and it appears to be a simple mixing operation.

However, the cost-effective operation of a modern blending plant is critically important to the overall process of delivering the correct lubricants of the correct quality and performance to customers. Blending lubricants may be relatively easy; operating a blending plant is certainly not.

7.2 LUBRICANT BLENDING AS PART OF THE SUPPLY CHAIN

The supply chain identifies the flow of information and material from a supplier to a customer and vice versa. This complex interaction should not be the responsibility of any one department. Everyone in the company, from managers downward, need to develop an understanding of how the new production methodologies can impact the supply chain, so that company-wide networking can be used.

Many blending plant operators do not yet understand the benefits of prioritising strategically planned *supply chain management*. When carefully planned and implemented, effective and efficient blending of lubricants can make a significant contribution to the company's overall profitability, through reducing operating and overhead expenses.

Product readiness and availability can be improved by proper forecasting of market demands, combined with a synchronised production schedule. Both are essential elements of an optimised supply chain. Product availability is not necessarily achieved through higher inventory, which only increases storage and handling expenses.

Plant expenditures will be impacted by reducing operation expenses, for both staff and equipment. Many companies operate on the Pareto principle. (An Italian-Swiss socio-economist and practicing engineer, Vilfredo Federico Domaso Pareto, observed that 20% of the people in Italy owned 80% of the wealth.) For companies, it is a general observation that 80% of the revenues (or profits) are accounted for by 20% of the products.

Unfortunately, 80% of the operating expenses are also accounted for by 20% of the products. By applying to this aspect of the business a correctly implemented “supply chain”, the process could become routine and extensively automated.

Reducing operation expenses is a major factor in good supply chain management.

7.3 KEY COMPONENTS OF A MODERN LUBRICANT BLENDING PLANT

7.3.1 Automatic Batch Blender

A typical automatic batch blender (ABB) system, a diagram of which is shown in Figure 7.1, has a stirred and heated blending vessel standing on load cells, with another, smaller weighing container to achieve the required precision for components added in small amounts.

Modern ABBs are usually made of stainless steel. The blending vessel is mounted on three load cells to allow the main components to be dosed by weight. The blending vessel is equipped with an efficient mixing system, suitable for the physical properties and formulation specifications of the lubricants to be blended. The mixing system can typically operate with a minimum level corresponding to 10% of the vessel’s capacity. Mixing units can be either paddle, agitator or dispersing type. Air sparging mixing is not recommended for blending lubricants. The reasons for this are discussed in Chapter 8.

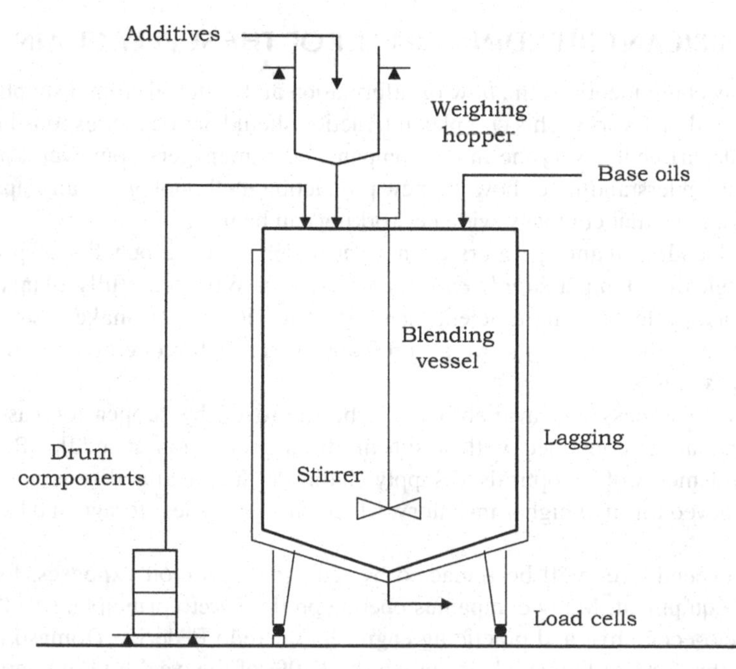

FIGURE 7.1 Diagram of an automatic batch blender.

The blending vessel can be equipped with external heating systems, such as a double jacket or welded half-pipe coils. Heat is optimally provided by low-pressure steam or hot oil. Some manufacturers of ABBs split the heating system into different zones which can be set up automatically to only heat the part of the blending vessel that contains base oil(s) and additives. This minimizes heating of the non-covered vessel wall and can help to maintain blended product quality.

Cross-contamination between blends and minimising the generation of "slop oil" (see also Section 7.3.7) is achieved when the blending vessel has a conical sloping bottom, with the offtake pipeline at the lowest point of the vessel's bottom. Ideally, the blending vessel should have a "pump-around" system, whereby the blend can be circulated around the vessel separately from the mixing unit, if operating conditions or blending times require this.

An ABB is based on weighing each component as it is added to the blend vessel and then stirring all the components together. The system is most suitable for batch sizes between about 2 and 20 metric tonnes, depending on the capacity of the blend vessel. Some blending plants have one or two 5-tonne ABBs and one or two 20-tonne ABBs.

An ABB that uses the most accurate load cells has the following advantages:

- Formulations are based on weight and can be processed automatically.
- The density, viscosity, pressure and temperature of the components have no effect on the dosing or blending procedure.
- The presence of air bubbles in the components has no effect.
- The set-up time is short.
- Batch sizes can be varied depending on production requirements.
- There is no slop oil.
- It is very easy to use with modern, complex formulations.

Mixing can begin as soon as the mixing unit is covered by liquid (usually the first base oil), so the product can be homogenised shortly after dosing the last component. The operator can manually add smaller components to the weighing container. The finished product is then pumped directly, via piggable pipelines (see Section 7.3.7), into holding tanks, analysed and filled in finished product units via filling lines.

The small weighing container used to dose minor components can be an automatic mini batch blender (AMBB), which uses the same technology as the ABB, but can also be used separately for producing smaller batches, for example, less than 5 metric tonnes. Liquids from the AMBB can be piped directly to the ABB or can be piped alternatively, again using piggable pipelines to product holding tanks.

A photograph of the top of a typical ABB is shown in Figure 7.2, and a photograph of the bottom of the same ABB is shown in Figure 7.3. Note that the load cells on this ABB are attached to the sides of the unit, not at the bottom as shown in Figure 7.1. Either an ABB can stand on the load cells, as shown in Figure 7.1, or the load cells can support the weight of the ABB and its contents attached to the sides of the ABB, as shown in Figure 7.4.

FIGURE 7.2 Photograph of the top of an automatic batch blender. (From Fluid Solutions GmbH. With permission.)

FIGURE 7.3 Photograph of the bottom of an automatic batch blender. (From Fluid Solutions GmbH. With permission.)

FIGURE 7.4 Photograph of a load cell installation of an automatic batch blender. (From Fluid Solutions GmbH. With permission.)

7.3.2 In-Line Blender

When the economic batch size is more than 50 metric tonnes, the use of an in-line blending unit is recommended. A process flow diagram for an ILB is shown in Figure 7.5.

With an in-line blender (ILB), between five and eight components (base oils and additives) are simultaneously heated and pumped through mass flow meters to a mixing unit and the mixture flows to a product tank. Smaller quantities of additives are added into the blender as drum components or via cocktail tanks with dosing pumps. When the required quantity is achieved, all inflow valves are closed and the batch is completed. For the production of large volumes of finished lubricants, particularly automotive engine oils, an ILB can be run continuously, for several days if necessary.

Previously, volume counters and mechanical regulators were used. Now, mass flow meters (generally micromotion Coriolis meters), software regulators based on fast programmable logic controller (PLC) systems and progressive rule strategies, are used in order to achieve high-quality products. This technique is also appropriate for difficult liquids, such as high-viscosity additives.

Modern ILBs have production capacities that range from 5 to 1000 metric tonnes, with a dosing accuracy of $\pm 0.05\%$. Although an ILB can be used for small batches, they are most cost-effective when used for larger-scale production.

Suppliers of modern ILBs will provide a blending plant with process engineering, equipment supply procurement, project management, fabrication, testing, commissioning, training, lifetime services and support and performance guarantees. A turnkey system will include flow meters, piping, valves, headers, instrumentation, mixers, pumps, wiring and the control system.

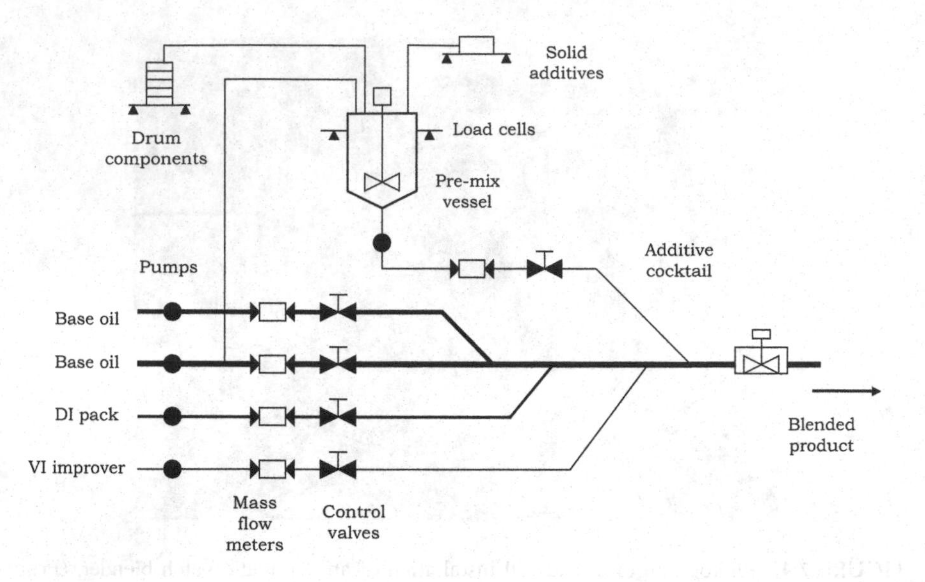

FIGURE 7.5 Process diagram of an in-line blender.

The advantages of an ILB are:

- Good suitability for the production of large batches.
- Thc ability to achieve an on specification finished product from blend header and to eliminate the need for mixing tanks during continuous blending.
- Production of instantaneous finished products via the static mixer in the blend header combined with ratio control.
- The ability to take samples during the production run.
- Achieving on-specification blending with real-time adjustments by the control system to correct errors in blend ratios or formula changes.
- The ability to accommodate premixes or parts per million additive additions for non-bulk additive dosing into the blend system.
- The product can be filled directly into holding (storage) tanks.
- The tank farm can be minimised.

Many suppliers of ILBs now offer systems that include wireless communication with the blending plant's distributed control system (DCS) (see Section 7.6). This can save on installation costs when upgrading an existing blending plant.

A photograph of a typical skid-mounted ILB is shown in Figure 7.6.

In order to maintain the quality of blended lubricants, the mass flow meters require regular recalibration. Most manufacturers of flow meters maintain their own facilities for their initial calibration before they are shipped. Many also offer recalibration services to their own customers, and some offer these services more widely. Three main methods are used for calibrating and recalibrating flow meters: master meter, piston prover and gravimetric.

FIGURE 7.6 Photograph of a skid-mounted in-line blender. (From Fluid Solutions GmbH. With permission.)

Another highly reliable method of calibrating liquid flow meters is to weigh the liquid that passes through the meter under test in a specified period of time, such as 1 minute. In order to obtain an accurate reading of the weight, a calibrated weighing scale is generally used. Actually implementing this method requires a little creativity. Since this is a timed measurement, the piping needs to be filled with flowing liquid before the test begins. At this point, the flow should be going into a discharge container. The flow can then be switched, by valve, into a vessel on the weighing scale, and switched back at the end of the test time.

End users who need to have their flow meters recalibrated have several choices of where to send their meter: an independent calibration facility, the flow meter manufacturer or a mobile service that does the recalibration on site. Variables to consider include where the flow meter manufacturer and the independent calibration company are located (which is closer), the turnaround time, the type of flow meter and the relative costs of recalibration.

Some companies offer mobile recalibration services that enable flow meter recalibration on site. Mobile calibration has the advantage of being able to most closely duplicate the actual operating conditions of the meter during the recalibration process. These services work especially well with smaller meters for liquid applications, including petroleum liquids.

In the author's experience, mass flow meters used in lubricant blending plants should be recalibrated every 3–6 months.

7.3.3 Simultaneous Metering Blender

With economic batch sizes between 20 and 70 metric tonnes, the use of a simultaneous metering blender (SMB) could be considered. A process flow diagram for an SMB is shown in Figure 7.7.

An SMB combines the volume production capacity of an ILB with the operating flexibility of an ABB. The SMB unit adopts the flow measuring technologies of an ILB, but instead of blending in a kettle, the components are sent through a header into designated storage tanks for mixing. The SMB blending unit is the simplest oil-blending unit and most appropriate for less automated blending plants.

The SMB is designed to simultaneously measure liquids with mass flow meters in the correct ratios through a header. The SMB unit contains a product feed pipe network, optionally separated by family groups (base oils and additives) and a piggable mixing pipe.

The technical design of the SMB unit consists mainly of two main differences compared with ABBs and ILBs. The SMB unit does not require a kettle and the output is transferred into a storage tank, equipped with mixers, for final homogenisation after pumping all the components into the tank.

Unlike an ILB, the components are dosed via the mass flow meters, according to the formulation, directly into a storage/mixing tank. This procedure has the advantage that a big batch can be dosed quickly into the homogenisation tank, using fewer mass flow meters, pumps and pipes. As the product remains in a homogenisation tank, it is possible to correct the product with later additional dosing at any time. After dosing all the additives, a final weight of base oil (to complete the formulation) is used to flush all the input pipes, thereby cleaning the SMB. The production of slop

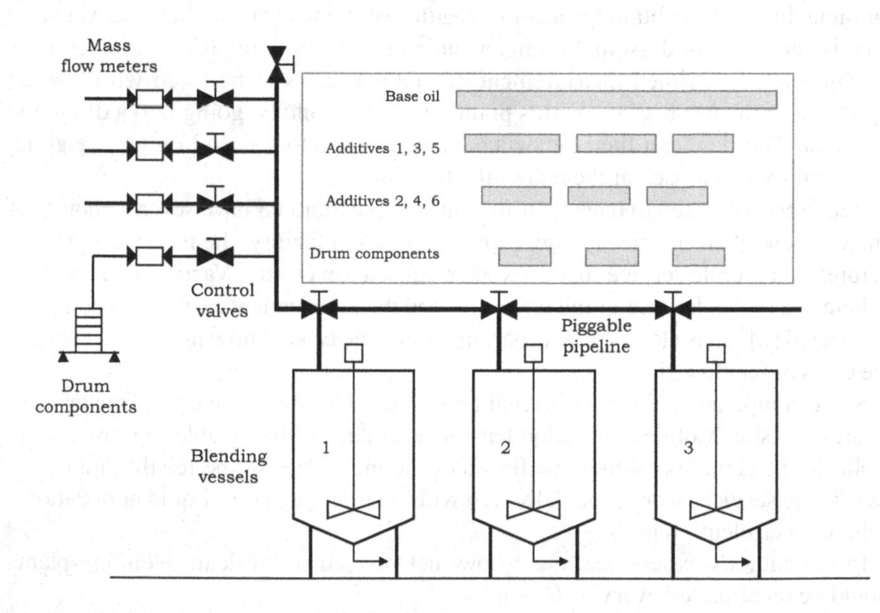

FIGURE 7.7 Process diagram of a simultaneous metering blender.

oil is thus avoided. Like an ABB, the storage/mixing tanks used with an SMB should have conical sloping bottoms and the offtake pipe at the lowest point in the bottom.

One SMB unit can be connected to any number of storage/mixing tanks. One storage/mixing tank can be dosed while another is circulated to finalise mixing, the contents of another tank can be tested for quality control and the contents of yet another tank can be transferred to a filling line.

The advantages of an SMB are:

- Good suitability for the production of large batches.
- Mass flow measurement, so formulations provided in kilograms can be processed directly.
- Efficient for any blend size, limited only to the size of the destination tank.
- Accuracy is not limited by batch size.
- A short set-up time, because only the drum line set-up time is required.
- It is possible to correct the product later, if necessary, in the blending tank.
- Simultaneous product loading speeds blending.
- Can be operated by the DCS formulation management system.
- Less mechanical engineering.
- Minimum contamination with pigging and flushing.
- No production of slop oil.

Some manufacturers of SMBs offer customised skid-mounted SMB units tailored to each customer's needs. These systems follow sub-assembly design principles, which enable cost-effective extensions and reduce maintenance costs.

A photograph of a typical SMB is shown in Figure 7.8.

FIGURE 7.8 Photograph of a simultaneous metering blender. (From Fluid Solutions GmbH. With permission.)

7.3.4 Drum Decanting Unit

Additives can be delivered to blending plants in bulk, in drums, in small containers or in bags (as powders). Transferring additives in bulk to either ABBs or SMBs is straightforward. Handling additives in small containers or as powders is also relatively easy. Completely emptying drums, particularly if the additive is viscous, can be difficult. Leaving even a few kilograms of additive in the bottom of a drum is wasteful and potentially costly. Every year, 5,000–10,000 drums of additives are used in a typical lubricant blending plant. If only 2 kg of additive is left in a 180 kg drum, valued at approximately US$5, the value of the unused additives could be as much as US$50,000 per year, so the cost of a drum decanting unit (DDU) can be recovered very quickly.

Using a manual or an automatic DDU, emptying drums is possible within a very short time. Integrated cleaning cycles as part of the formulation enable complete discharge of expensive additives while avoiding the production of slop oil. A DDU system is designed for adding a required amount of liquid from drums to blending vessels or mixing headers in an efficient and accurate way, without causing contamination between the different drums. A rinsing cycle washes out remaining additives from an empty drum and leaves a washed drum without producing any slop oil.

The system consists of a weighing platform for accurate measuring, a rinsing kettle containing hot base oil for cleaning of a lance and drums and a lance positioning system that controls the vertical and horizontal movements of the lance. Operating a DDU, a diagrammatic representation of which is shown in Figure 7.9, is relatively easy, as discussed in Section 7.5.3.

A photograph of a typical DDU is shown in Figure 7.10.

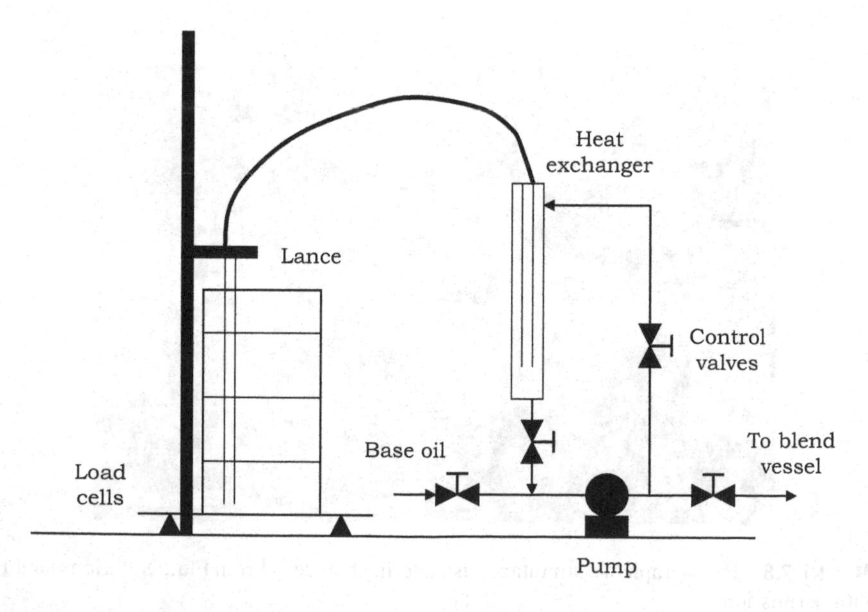

FIGURE 7.9 Process diagram of a drum decanting unit.

FIGURE 7.10 Photograph of a drum decanting unit. (From Fluid Solutions GmbH. With permission.)

7.3.5 Premix Blending Vessel

A premix blending vessel is, in essence, a small batch blending vessel, which is operated either manually or automatically, using load cells. The capacity of a premix blending vessel is usually only 1 or 2 tonnes. These vessels are almost always associated with an ILB or an SMB, although a number of premix vessels can be located in a separate area of the blending room, to be used to feed either an ILB or an SMB. An ABB usually has its own premix blending vessel, as illustrated in Figure 7.1. Premix blending vessels are almost always heated and lagged.

Premix blending vessels are used to dissolve solid additives or to reduce the viscosity of viscosity index improver liquid concentrates, prior to blending. They can also be used to mix a number of additives in small amounts, to prepare a "cocktail" of additives, again prior to blending. Obviously, the weights of components, including the base oils, need to be predetermined, as part of the master blend in either the ILB or the SMB.

If an ILB is being operated for a long period (more than 5–7 days) or continuously, two premix blending vessels may need to be used, to enable this operation. One premix vessel will be feeding the additive mix to the ILB, while the other premix vessel is being used to prepare the same additive mix or "cocktail", ready for use when the contents of the first vessel have been used.

7.3.6 Drum Heating Unit

Blending of base oils and additives is usually done at temperatures ranging from 40°C to 60°C. Heating base oils and additives in bulk storage tanks prior to blending and heating blending vessels is simple, using either low-pressure steam or hot oil pipes. Electric heaters used in either storage tanks or blending vessels, even those suitably controlled to avoid overheating, are not recommended for a lubricant blending plant.

FIGURE 7.11 Individual drum heating unit.

Heating additives in drums requires either individual drum heaters (usually electric) or a heating cabinet or heated room in which drums can be placed for 24 hours prior to decanting using the DDU. Individual drum heaters can be either a cylindrical insulating jacket that surrounds the drum and contains a heating element or a cylindrical unit that folds open and closed, with a folding lid, that contains either two heating elements, one on each half cylinder, or one heating element at the base of the unit.

Insulating jackets with an internal heating element can also be obtained for intermediate bulk containers (IBCs). Again, the temperatures for drum heating units need to be carefully controlled, to avoid overheating the additives, which may cause their decomposition prior to blending.

A typical individual opening cylinder drum heating unit is shown in Figure 7.11, a drum heating cabinet is shown in Figure 7.12 and a typical heating jacket for an IBC is shown in Figure 7.13.

7.3.7 Pigging Equipment

As noted in Chapter 6, lubricant blending plants face increasing demands for flexibility, higher production and lower costs. One of the factors that can impact all three of these is the cleaning of the product transfer pipelines. In a modern blending plant, the installations are generally designed as "multiple product/multiple pipeline plants".

In the past, the only way that an operator of a batch process in a lubricant blending plant could ensure that a product was completely cleared from a pipeline was to flush it with a cleaning agent, such as a lower-viscosity base oil. This procedure generated a "waste" that came to be known as "slop oil". In practice, some of the slop oil was frequently used either in lower-performance lubricants or in the manufacture of greases. Any slop oil that was not used in this way had to be either used as a fuel or sent to a used lubricant re-refinery for reprocessing. In all cases, the generation

FIGURE 7.12 Drum heating cabinet. (From AMARC srl. With permission.)

FIGURE 7.13 Intermediate bulk container heating jacket.

of slop oil added to blending plant costs, reduced production volumes and lowered blending plant flexibility.

An alternative method of clearing a pipeline is to use a "pig", an inflatable rubber ball that can be blown through the pipeline using compressed air, to push all the liquid in front of it to the holding tank.

Pigs have been used for many years in the oil industry, to help clean and maintain crude oil and fuel products pipelines. Some early cleaning pigs were made from straw bales wrapped in barbed wire, while others used leather. Both made a squealing noise while travelling through the pipeline, sounding to some like a pig squealing, which gave pigs their name.

In oil product manufacturing, pigging can be used for almost any section of the transfer process between, for example, blending, storage or filling systems. Pigs are used in lubricant blending to clean the pipes to avoid cross-contamination and to empty the pipes into the product tanks. Usually, pigging is done at the beginning and end of each batch, but sometimes it is done in the midst of a batch, such as when producing a premix that will be used as an intermediate component.

A major advantage for multiproduct pipelines of piggable systems is the potential of product savings. At the end of each product transfer, it is possible to clear out the entire line contents with the pig, usually forward to the receipt point, or sometimes backward to the source tank. There is no requirement for extensive line flushing. As a result, there is almost no generation of slop oil. (Very small volumes of product may be left clinging to the inside walls of pipelines.)

Without the need for pipeline flushing, pigging offers the additional advantage of much more rapid and reliable product changeover. Product sampling at the receipt point is faster with pigs, because the interface between products is very clear. The old method of checking at intervals to determine whether the product is on specification takes considerably longer. Pigging can also be operated completely automatically by a PLC. Pigging has a significant role to play in reducing the environmental impact of batch operations. All these problems can now be eliminated due to the very precise interface produced by modern pigging systems.

If the pipeline contains butterfly valves, or reduced port ball valves, the pipeline cannot be pigged. Full-port (or full-bore) ball valves cause no problems because the inside diameter of the ball opening is the same as that of the pipe.

Pigging equipment comprises pig launchers, pig receivers, an air compressor and pipework to get the compressed air to the launchers and receivers. This is illustrated in Figure 7.14.

During blending, pigs are held in a "parking position" immediately below an outlet valve and pipeline in each of the blending vessels. When blending has been completed, a holding tank is selected and the inlet valve to that tank is opened; the pipeline that connects the blending vessel and the holding tank has been cleared previously during the last pigging operation.

The blending vessel outlet valve is opened and the blend is pumped to the holding tank. When the blending vessel is empty, the blending vessel outlet valve is closed and the pig is "launched" into the transfer line, using compressed air.

Using compressed air, the pig is pushed all the way along the transfer pipeline, pushing the remaining product before it and into the holding tank. Note that all the other valves in the transfer pipeline are closed, so that product can only flow into the designated holding tank. When the transfer pipeline has been cleared and the pig has arrived at a "receiving position", the holding tanks' inlet valve is closed and the pig can then be returned to its "parking position", again using compressed air. With modern designs of piping in blending plants, pigs can successfully negotiate rounded bends in transfer pipelines. A photograph of a pigging valve is shown in Figure 7.15, and a photograph of the piggable discharge pipeline from an ILB is shown in Figure 7.16.

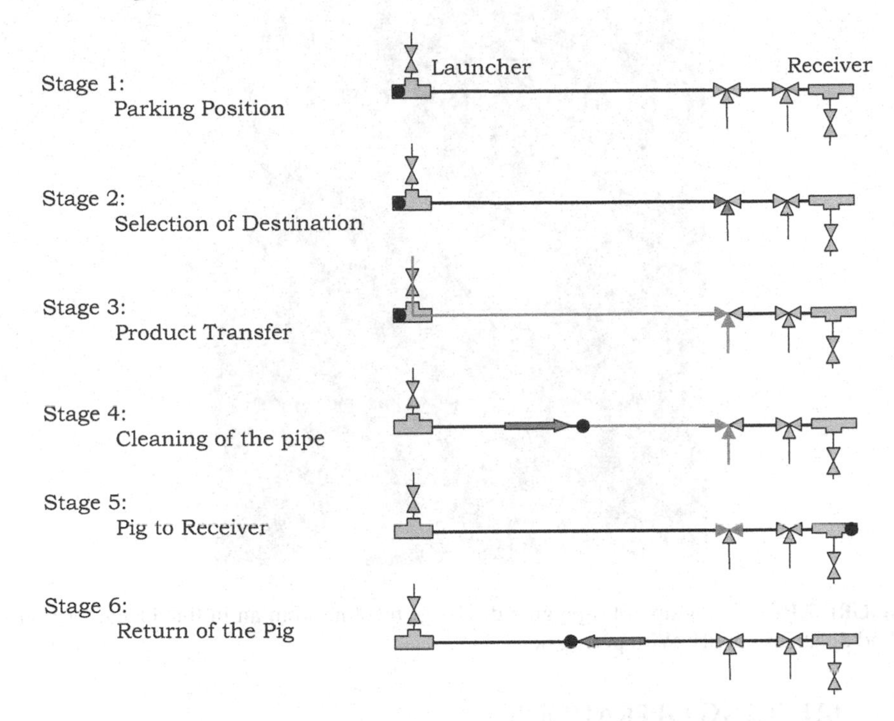

FIGURE 7.14 Operation of pigging equipment.

FIGURE 7.15 Photograph of a pigging valve in an in-line blender. (From Fluid Solutions GmbH. With permission.)

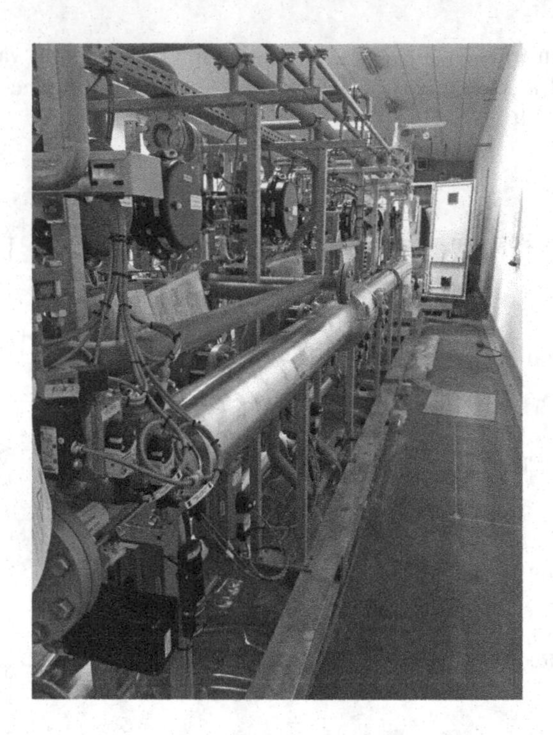

FIGURE 7.16 Photograph of a pigging discharge pipeline from an in-line blender. (From Fluid Solutions GmbH. With permission.)

7.4 BLENDING OPERATIONS

7.4.1 Batch Blending

In batch blending, the raw materials (base oils and additives) are placed in a large vessel or blending tank (the ABB), where they are mixed until homogeneous. They are then transferred to intermediate storage and a new batch is commenced.

In the past, a common design of the batch blending unit the so-called "vertical" design in which the blending vessel is located on a floor above the main holding tankage, into which the product can flow by gravity. There is another floor above the blending vessel, from which raw materials are loaded into the blending vessel. This type of plant requires a strong and relatively expensive building to contain the blending vessel, and there is often little space on the upper floor to contain materials such as additives, all of which have to be raised to this loading floor.

With improvements to the capacity and reliability of pumps, the horizontal type of plant is now favoured, provided that there is sufficient space to accommodate it economically. In this type of plant, materials are pumped laterally from one place to another, and working space is less restricted. In all plants where pipelines are used for more than one product, provision for pipeline flushing or pigging must be provided.

In either case, base oil supply is required in the top of the blending vessel, usually by a dedicated pipeline from a bulk storage tank. Blending agents and additives are normally provided from bulk supply only in large and particularly in continuous plants. Additives are more normally taken from drums, using the DDU and a small transfer pump.

For some types of lubricants, solid additives are used, and these will be unloaded from bags into hoppers on the floor above the blending vessel. Some viscosity index improvers are available as rubber crumb, in which case this would be treated in the same manner, but care must be taken to avoid agglomeration. For piped products in cooler climates, outflow or suction heaters on the bulk storage tanks are desirable, and the tanks themselves may be required to be heated to keep the contents at a reasonable temperature.

In the case of additives, it is important that only low-pressure steam, hot water or hot oil is used for heating, as excessive temperatures can lead to decomposition of many additives. Heating coils must always be sound and inspected from time to time because water can rapidly accelerate additive decomposition and, if carried from a base oil tank into the blending vessel, can ruin a blend.

Hot oil is increasingly being used in heating coils to eliminate internal corrosion and to minimise product loss if leaks do occur. In the case of drummed additives, it will be necessary to provide either a drum oven or electrically heated jackets which can be used to heat the contents of the drums to a temperature at which they will flow easily. Again, care must be taken particularly with heating jackets not to cause local overheating of the additives.

To make a blend, in the case of simple plants, about 80% of the base oils are added to the blending vessel, which must be clean or contain only a heel of a similar oil from the previous blend. The base oils are heated to around 60°C, and then the additives are added. If there are no very viscous additives, the temperature can be lower and these may be added in the order of increasing viscosity, the blend being agitated throughout the addition process. In the case of some viscous viscosity modifiers, it may be desirable to heat the base oil to approximately 80°C, add the viscosity modifier and mix well, and then cool the batch before adding the more temperature-sensitive detergent additives. For some sensitive blends (such as automatic transmission fluids), there may be a preferred order of component addition at specified temperature maxima, and in these cases the advice of the additive supplier should be followed. In all cases, great care must be taken not to overheat additive components.

The measurement of the components can be either by volume or by mass/weight. Weight is now generally used, with the blending vessel on load cells, or alternatively, the base oils may be measured by hydrostatic gauge and the additives weighed by difference on a drum scale. The proportions of additives and base oils are, of course, different depending on whether the measurements are to be by volume or by weight, and it is important that the correct figures are used.

Mass flow meters (Coriolis meters) are available which can measure mass flow through a pipe, in which case proportions are as for weight.

It is important that the ingredients are thoroughly mixed together and that the blend is uniform. The simplest method, applicable to medium-sized blends, is to agitate the blend by blowing in compressed air through a perforated pipe, sparge ring or jet nozzle at the bottom of the tank. A more sophisticated version of this is the "Pulsair" system, in which a programmed series of air bubbles provides efficient agitation. However, the use of compressed air for mixing is not recommended for lubricants, because of the risk of oxidising sensitive additives. This is discussed further in Chapter 8.

The most common method is to have some form of mechanical agitator, which is usually some type of propeller mixer. For making small blends of solid additives in oil, a high-speed turbine mixer (Silverson mixer) is often employed. Mixing can be assisted by a pump-around system, which means that product is withdrawn from the side or bottom of the tank and pumped back into the top, providing a constant circulation of product. A variation of this, which is an efficient blending system in its own right when set up correctly, is to pump the material back through a jet nozzle placed in the bottom of the tank, which agitates the product vigorously when the pump-around system is switched on.

The length of time required for mixing will depend on the agitation system and the viscosity of the blend. It will almost always be judged by previous experience. For new and possibly difficult blends, it is advisable to turn off the agitation and, after the product has settled, to compare the appearance and possibly the viscosity of top and bottom samples from the blending vessel. At this stage, the blend is not completed because not all the base oil may have been added at the beginning of the process. Assuming that the full amount of each additive required to produce the full volume of oil was programmed, it will be necessary to add the remaining base oil. The reason for holding this back is that the viscosity of the finished blend must lie within certain limits, and there will always be slight differences in individual batches of both base oil and additives, in terms of their viscosities. It is therefore necessary to measure the viscosity of the semi-finished blend and to calculate the adjustments needed in terms of heavy and light base oils to meet the target viscosities. Calculated correction amounts of base oil can be mixed in fairly rapidly, after which the blend viscosity should be checked again and possibly a last adjustment made. The amount of adjustment necessary, and therefore the quantity of oil held back, can be minimised by experience and by knowledge of the actual viscosities of the components used in the blend and how these differ from nominal values.

When it is sure that the blend is homogeneous and on grade for viscosity, if it cannot be held in the blending vessel it should be transferred to a holding tank from which samples are sent to the laboratory for full release testing. If this testing indicates that the blend is either too rich or too lean in additive content, then at some later stage it will have to be pumped back into the blending vessel and the blend adjusted once again. If there is a large holding tank and it is required to consolidate several blend batches into this tank, then they should all be blended to meet the specification, but space left in the tank for a final adjustment batch. The holding tank is mixed before testing, and if this is found to be off grade, then the final batch should be blended to correct the contents of the holding tank and bring the whole tank on specification.

When the contents of the holding tank are released as being on specification, the lubricant can then be passed to the filling pipeline where, after passing through a filter, drums, bottles or cans are filled with the product. Filling is discussed in Chapter 12. Of course, it is important that the product does not get contaminated in the filling system. This should therefore be thoroughly drained between products and preferably pigged or flushed with a small quantity of base oil. Drainings from washings of lines and the blending vessel and other tankage are often used as fuel, but it is possible with care to segregate them and incorporate them into future blends of the same type. Most automotive engine oils can be considered the same type of oil, and these must be separated from gear oils and from industrial oils. In many blending plants, separate blending vessels are used for exclusively blending certain types of oils, in order to avoid cross-contamination. This is discussed in Chapter 8.

7.4.2 Automated and In-Line Blending

For the production of many grades of oils, it is possible to automate batch blending. This can be achieved by having the mixing vessel on load cells and introducing all the raw materials in the form of liquids which can be dosed by pipeline. This may mean that the more viscous additives need to be cut back with suitable base oil either by the additive supplier or on site.

Because the number of incoming pipelines is limited, in the case of complex blends, such as when oils are being blended from component additives rather than packages, it may be necessary to combine some additive streams by premixing. The addition of both base oils and additives to the mixing vessels can be computer controlled and a continuous stream of consistently on-grade product produced.

A more elaborate scheme has used small preblend vessels like robots, which move around tracks to different filling heads, their contents being mixed together either by tank mixing or by in-line homogenisation. The whole operation is controlled by a computer.

Another type of automated plant is the proportional blender or ILB. This is used by many oil companies for producing the very large volume of automotive engine oils and certain industrial oils which constitute a major part of their business. As for the automated batch blending process, all components have to be available as non-viscous liquids, which frequently means cutting back on site with a base oil which is a component of the blend. This may be done in small batch-type blenders.

A limited number of components are metered into a mixing line with their rates of flow adjusted so that all products are continually metered in quantities designed to finish up with the correct blend mixture. This of course is achieved by computer control. In the mixing line, a vortex is created and the various input streams are mixed so that they emerge at the end of the line as a virtually homogeneous blend. This is pumped to a holding tank where usually a final mix is performed by a pump-around system, and the product is sent for quality control testing.

For a complex blend, if there are insufficient entry ports into the mixing line for all the components of a blend, some of these will need to be combined in the preblend tankage used for cutting back the viscous additives. In this way, all the components can be continuously and simultaneously dosed.

The system is ideal for a relatively small number of very large-volume blends, and once set up it can produce continuously on-grade product. As relatively high mixing energy is applied to blends, checks should be made that any viscosity index improving or pour point depressant polymers used do not suffer unacceptable levels of shear breakdown.

For an ILB, the blending temperature is set by the inlet temperatures of each of the charged raw materials. For this reason, it is advisable to have each of the raw materials at about the same temperature, so that there is less risk of layering through the vortex mixer. The ideal blending temperature will be set by the viscosity of the most viscous raw material or mixture, making sure that the temperature is not too high, thereby risking degradation of one or more of the additives. The more viscous raw materials may need to be cut back with lower-viscosity base oil.

7.4.3 Operating a Drum Decanting Unit

A DDU is designed to automate the process of decanting high-value additives from drums or IBCs into a blending vessel. A DDU is usually made of stainless steel and consists of a drum platform incorporating a roller conveyor, a tilting mechanism and a weighing scale, a lance, a rinsing kettle, a decanting pump and a rinsing pump. The rinsing kettle can have an optional weighing sensor on mounting legs.

Preprogrammed formulations can be selected using a full-colour touchscreen, and on-screen instructions can guide the operator safely through each stage of the process. When nearing empty, the drum can be tilted to ensure that maximum product removal is achieved.

A drum is placed onto the weighing scale underneath the lance. The operation is started, which activates the filling cycle, and the lance is then lowered into the drum. The pump removes a predetermined amount of liquid from the drum and pumps it to a blending vessel. The lance exits the drum and enters the lance cleaning tank (rinsing kettle), where it cleans itself inside and out, ready for use on another product.

When a drum is nearly empty, the lance can be used to introduce hot oil into the drum. The hot oil solubilises the remaining additive and the mixture can then be pumped out of the drum to the blending vessel. The weight of hot oil and additive can be predetermined as part of the blend in the blending vessel. This helps to ensure that additive wastage and the generation of slop oil are minimised. Also, the rinsing kettle hot oil is used to flush the lance of any remaining product, to minimise the possibility of cross-contamination.

7.5 SAMPLING RAW MATERIALS AND BLENDS

7.5.1 Importance of Sampling

Lubricant sampling and extraction is perhaps the most important and highly variable step taken prior to the analysis of a sample. It is also the easiest to make consistent. A common mistake made by some blending plants is that lubricant sampling is not considered an important part of the operation. The methods and procedures used for sampling will determine the amount, accuracy, reliability and utility of the data that can be acquired from the sample.

Samples of base oils and additives need to be taken before blending, samples of blends may need to be taken during blending and samples of finished products will need to be taken following blending. It is very important that samples are taken while equipment is operating (whether blending or pumping), so that the sample is representative of the process being conducted. Equipment must never be started just to take a sample, as it will not be representative.

7.5.2 Sample Bottles

Step 1 for obtaining a truly representative sample of a base oil or liquid additive is to use the most appropriate sample bottle. Many companies believe, mistakenly, that any bottle or container in which to deliver the sample to a laboratory will be okay. It is imperative that sample bottles are clean (preferably ultra-clean), dry and free of any material that may contaminate the sample. Drawing a sample into a washed-out beverage bottle will not be good enough. Even a sample bottle purchased with the lid and bottle in separate packages will not be sufficient.

Oil sample bottles are available in a few standard materials, namely plastic or glass. The material should be selected based on the type of fluid sampled and the cleanliness requirements. The most common sample bottles are high-density polyethylene (HDPE) or polyethylene terephthalate (PET). HDPE is opaque, which may be its main disadvantage. Not having the ability to clearly see the oil in the bottle prevents visual analysis, which can be helpful in detecting water or heavy particle contamination. Conversely, PET is clear, but generally not suitable for samples at temperatures higher than about 90°C. However, PET has greater compatibility than HDPE with most industrial lubricants. Compared with glass bottles, both polyethylene-based bottles are relatively inexpensive, but they offer the benefits of excellent cleanliness levels and lubricant compatibility.

The size of the sample bottle should be based on the type of sample fluid, together with the number and type of tests to be conducted. For most standard oil analysis tests, oil samples are taken in a 100 or 120 ml bottle. For advanced or exceptional tests, a 200 ml or larger bottle may be required, although bottles larger than 200 ml tend to be used for fuel analysis.

The blending plant laboratory will provide advice on the size of sample, and therefore sample bottle, required for each type of base oil or additive sample, together with the cleanliness requirements. When purchasing sample bottles, or using sample bottles provided by an independent oil analysis laboratory, it is very important to know that they are cleaned to the specifications required to hit the target cleanliness goals.

7.5.3 Sampling Methods

Several methods can be used for obtaining oil samples. Of course, some are more effective than others; the aim is to make the method in which the base oil, additive, blend or product is sampled consistent. It is very important to ensure that each time a sample is drawn, the end result is the same regardless of the technician drawing the sample. Written procedures and specific training are vital to success.

Drop-tube sampling is an effective, low-cost way to draw a sample with a vacuum pump. A plastic tube is inserted into a storage tank, drum or blending vessel, and a vacuum draws the sample into a sample bottle. The method, however, has three major drawbacks. The sample may be taken from very variable locations inside the tank, drum or vessel, and so may not be consistent. Opening the tank, drum or vessel potentially allows significant amounts of airborne contamination to enter the oil or additive. A large volume of sample may need to be used to flush the sampling tube, unless a new tube is used for each sampling operation. Additionally, there may be problems with sampling high-viscosity fluids. In summary, this method of oil sampling should be avoided when other methods are available.

The most widely used implement for sampling a drum of base oil or liquid additive is a glass tube, commonly referred to as a "glass thief". This device is simple, cost-effective and quick and collects a sample without having to decontaminate. Glass thieves typically have an inside diameter of between 6 and 16 mm and are around 1.2 m in length. The tube is open at both ends, with the openings slightly narrower than the rest of the tube. The glass must be comparatively thick and robust.

Before sampling, the top of the selected sample bottle should be removed, as opening the sample bottle after the sample has been taken is likely to be quite difficult. The drum's bung can also be removed at this stage. (This should be done either indoors or when it is not raining or snowing outdoors.) The top of the glass thief is capped with a tapered stopper or thumb, ensuring that liquid does not come into contact with the stopper. The bottom end of the glass thief is inserted into the drum, keeping the top end capped, and the tube is lowered to around the middle of the drum. The stopper can then be removed, to allow the liquid in the drum to flow into the glass tube. The top of the glass thief is then recapped and the tube is removed from the drum, allowing any liquid adhering to the outside of the tube to drain back into the drum. When this is done, the bottom end of the tube is lowered into the sample bottle and the contents of the tube are allowed to drain into the sample bottle. The sample bottle's top can then be put back and the sample sent to the blending plant laboratory for analysis. The glass thief can then be cleaned and dried, ready for its next use.

Drop bottle sampling can be used for storage tanks. A prestoppered sample bottle (most usually either glass or PET) is lowered into the tank, from a hatch in the top of the tank, in a weighted cage. Once the cage has reached a predetermined height in the tank, a second rope is used to remove the stopper, so that the fluid sample can enter the bottle. The cage and bottle are then raised to the hatch and the bottle restoppered. This method is very satisfactory for obtaining top, middle and bottom samples of a tank's contents. However, for most base oils and additives, it can be very messy, requiring much cleaning of ropes, cages and the outsides of sample bottles.

A similar implement is known as a "bacon bomb sampler", shown in Figure 7.17. It consists of a cylindrical body, usually made of chrome-plated brass and bronze with an internal tapered plunger that acts as a valve to allow the sample to enter the sampler. A line (rope) attached to the top of the plunger opens and closes the valve.

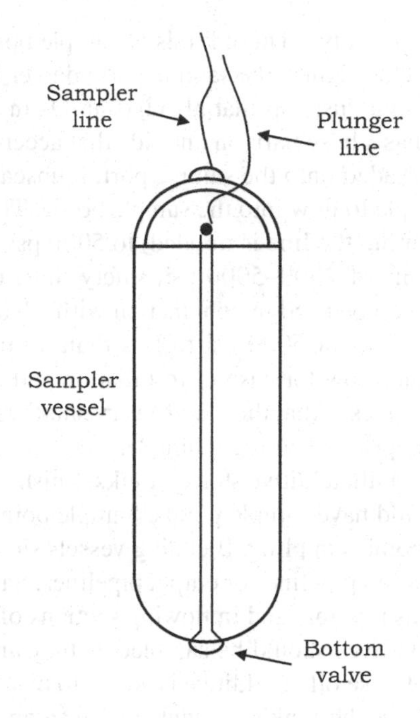

FIGURE 7.17 Diagram of a bacon bomb sampler.

A rope is attached to the removable top cover, which has a locking mechanism to keep the plunger closed after sampling.

To take a sample, the sample line and the plunger line are first attached to the sampler. Depending on whether the sample is to be taken from the upper, middle or lower third of the tank, and depending on the size and height of the tank, the sample line should be marked at the required depth. The sampler can then be lowered, using the sample line, until the desire depth is reached, when the plunger line is pulled to allow the sampler to fill. The plunger line is then released, to seal off the sampler. The sampler can then be pulled up, again using the sample line, being careful not to pull on the plunger line, so preventing accidental opening of the bottom valve. The outside of the sampler is then wiped, to remove oil adhering to it. The contents of the sampler are emptied into the sample bottle by pulling the plunger line, thereby opening the lower valve.

Drain port sampling can also be used for storage tanks and blending vessels. The location of the drain port is important, as a location that could collect debris at the bottom of the tank or vessel may give an unrepresentative sample. This may still be true even if high volumes of oil are flushed through the drain port before the sample is collected. A sampling point on the outlet pipeline from the tank or vessel is preferable to a drain port.

Several sampling valves are available commercially. Some valves are far superior to others. When selecting a valve, consider all the options. Sampling with valves as opposed to static sampling adds integrity and success to an operation. Perhaps the most effective option, which is typically used on larger, pressurised systems, is a

"Minimess" style of sample valve. These kinds of sample ports are check style; that is, the valve is normally closed until the sample port adapter is threaded on. Sample ports come equipped with a dust cap that also has an O-ring for second-stage leak protection. The adapter has a hose barb on one side that accepts standard plastic tubing. As the adapter is threaded onto the sample port, it unseats the check ball in the valve and allows the sample to flow into the sample bottle. These valves can be used on systems from 0 (assuming the line is flooded) to 5000 psi.

On pressurised systems of 2000–5000 psi, safety must be considered. Sample valves are available that can be used in conjunction with a handheld pressure-reducing valve to reduce pressures of 5000 psi to less than 50 psi. They are sold with several adapter styles that allow for ease of installation and use. Another benefit to these types of sampling valves is that they hold a very small volume of static oil. This results in less oil flushing prior to taking a sample.

Base oil storage tanks, bulk additive storage tanks, finished product storage tanks and all transfer lines should have sample points. Sample points on tanks should permit dip bottle or bacon bomb sampling. Blending vessels should have sample points in either or both recirculation pipelines or output pipelines. Sample points on transfer lines should be as short as possible and in flowing sections of the lines.

Base oils and bulk additives should be sampled as they are transferred into storage tanks. If a delivery of base oil or additive is added to material already in storage, it may be necessary to mix the tank's contents and resample if the new batch has different properties than the previous contents.

Drums or sacks of additives also need to be sampled. However, it is both expensive and impractical to take a sample from each drum or sack, unless there is only one of either in a delivery. With multiple drum or sack deliveries of additives, random sampling is likely to be satisfactory. The number of samples will depend on the number of drums or sacks.

A representative sample of each batch of blended lubricant must be taken, for process control, quality control and quality assurance purposes. This will be discussed in more depth in later chapters.

Samples of each delivery of base oils and additives and all batches of finished blends should be retained for at least 2 years.

7.6 AUTOMATION OF BLENDING

At present, to remain competitive in manufacturing lubricants generally requires high-performance automated systems, including blending, pigging, filling and storage. The automation of production plants has advanced continuously in recent years, using computerised control systems. In the past, with management information systems involving financial and administrative tasks in lubricant blending plants, the automation of blending was not part of the system. This made just-in-time (JIT) production increasingly difficult.

The aim of state-of-the-art automation systems is to view the production of lubricants as whole, in order to reach common solutions for the whole business. This guarantees the flow of information between the commercial production planning systems and the process control systems, which are oriented toward the production process.

Software tools have been developed over many years by a number of suppliers, often using proprietary software engineering. Initially, because of the variety and diversity of these tools, users of the control system in a blending plant incurred additional costs, caused by having to integrate blending programmes into subordinate operating systems.

During the last 20 years, the numbers of products, in addition to international, national and original equipment manufacturers' (OEMs) specifications required to meet customers' demands, have increased continually. Many customers now want more regular deliveries and shorter lead times. This means lubricant blending plants may need to manufacture smaller batches. Achieving quality targets (discussed in Chapters 11 and 14) can be difficult, due to poor process control, human error, cross-contamination or raw material variability. The ability of a lubricant blending plant to maximise the productivity of assets, achieve on-specification blends every time and maintain flexibility to respond to changing market demands is of paramount importance.

Modern automated batch management programmes provide powerful application software packages for configuring, scheduling and managing batch operations. They aim to improve batch production, consistency, traceability and profitability. Increased productivity tools provide the dexterity, speed and control needed to respond to increasing customer demands. They can be used to model, execute and track information associated with material and control flow across the lubricant blending plant. They are typically aligned with industry standards, such as ISA-88, ISA-95, IEC 61512 and IEC 62264. (IEC is the International Electrotechnical Commission.)

ISA 88, shorthand for ANSI/ISA-88 (American National Standards Institute and International Society of Automation), is a standard that addresses batch process control. It is a design philosophy for describing equipment and procedures. It is not a standard for software because it is equally applicable to manual processes. It was approved by the ISA in 1995 and is updated regularly. The last update was in 2010. Its original version was adopted by the IEC in 1997 as IEC 61512–1.

Currently, ISA-88 has four parts:

- ANSI/ISA-88.01–2010: Batch Control Part 1: Models and Terminology.
- ANSI/ISA-88.00.02–2001 Batch Control Part 2: Data Structures and Guidelines for Languages.
- ANSI/ISA-88.00.03–2003 Batch Control Part 3: General and Site Recipe Models and Representation.
- ANSI/ISA-88.00.04–2006 Batch Control Part 4: Batch Production Records.

The standard provides a consistent set of guidelines and terminology for batch control and defines the physical model, procedures and recipes. The standard sought to address a lack of a universal model for batch control, the difficulties in communicating user requirements, common procedures for suppliers of batch automation systems and difficulties in batch control configuration. The standard defines a process model, a process which consists of an ordered set of process stages, which consist of an ordered set of process operations, which consist of an ordered set of process actions.

ISA-95 (ANSI/ISA-95) is an international standard for developing an automated interface between enterprise and control systems. This standard has been developed for global manufacturers. It was developed to be applied in all industries and in all types of processes, including batch, continuous and repetitive processes. The objectives of ISA-95 are to provide consistent terminology for communications between system suppliers and process users, to provide consistent information models and to provide consistent operations models, which are foundations for clarifying application functionality and how information is to be used.

There are five parts of the ISA-95 standard:

- ANSI/ISA-95.00.01–2010: Enterprise-Control System Integration: Part 1: Models and Terminology.
- ANSI/ISA-95.00.02–2010: Enterprise-Control System Integration: Part 2: Object Model Attributes.
- ANSI/ISA-95.00.03–2013: Enterprise-Control System Integration: Part 3: Activity Models of Manufacturing Operations Management.
- ANSI/ISA-95.00.04–2012: Enterprise-Control System Integration: Part 4: Objects and Attributes for Manufacturing Operations Management Integration.
- ANSI/ISA-95.00.05–2013: Enterprise-Control System Integration: Part 5: Business-to-Manufacturing Transactions.

At the time of writing, Parts 4 and 5 of ISA-95 were still being developed.

The IEC 62264 standard is based on ANSI/ISA-95. The standard has six parts, each of which are contained in separate IEC documents:

- IEC 62264–1:2013: Object Models and Attributes of Manufacturing Operations (first edition 2003).
- IEC 62264–2:2013: Object Model Attributes (first edition 2004–2007).
- IEC 62264–3:2016: Activity Models of Manufacturing Operations Management (first edition 2007–2006).
- IEC 62264–4:2015: Objects Models Attributes for Manufacturing Operations Management Integration.
- IEC 62264–5:2016: Business-to-Manufacturing Transactions.
- IEC 62264-6:2016: Messaging Service Model.

Further details of these standards can be obtained from the International Society of Automation (www.isa.org), the American National Standards Institute (www.ansi.org), the International Electrotechnical Commission (www.iec.ch) and the International Organization for Standardization (ISO) (www.iso.org).

Suppliers of batch management systems generally claim that they offer increased product consistency, resulting in better quality, easy-to-use formulation management functions that reduce lead times for customers, integrated production management and control for maximum equipment utilisation and lower operating costs, together with compliance with regulations, by using embedded system technical features.

All modern batch management systems are able to sequence the process when steady-state conditions persist and nothing goes wrong with either the equipment or the process. Some newer programmes are designed to help recover from unexpected events, by supporting dynamic recipe parameters. Master formulation procedures can be configured with one or more formulation parameters defined as mathematical expressions, rather than as constant values. This enables the control system to evaluate one or more variable process conditions and dynamically update the formulation parameters in real time during operations. Improved product quality and reduced processing time are claimed to be two of the potential benefits when using dynamic parameters in control recipes.

Often, operator or supervisor intervention in processing a batch means aborting the batch recipe procedure and manually completing the batch. Newer batch management programmes provide tools to allow operator intervention to make the necessary adjustments and continue the processing of the batch through the control recipe procedure. An operator or supervisor can easily reroute the batch path to another allowable blending vessel if the originally selected vessel is not available or out of service. The most powerful feature, however, is an extensive runtime editing capability. Operators with the correct permission level can perform a runtime edit on any currently executing procedure. The runtime edit feature allows the modification of any subsequent operation or phase in the operation being executed currently, without stopping the operation. Of course, all operator-initiated changes are captured in the system audit trail and batch production record.

All process automation systems support interlocking strategies at the control module level. Interlocks required for safety and for equipment protection need to reside at the controller level. However, in a flexible batch production facility there could be one or more conditions that are dependent on the type of material being processed in the unit. High-performance systems can provide exception procedures as an extension of the ISA-88 procedural model, to provide the ability to configure recipe-specific error handling logic. An exception procedure monitors for undesirable product-related process conditions and contains corrective measures in case the conditions occur.

The most modern automated batch management programmes include a number of main functions:

- Definition management: The master batch recipe procedure and all nested procedure levels are configured graphically, with procedure, formulation, equipment requirements and other information specified for each individual product formulation.
- Execution management: The simultaneous execution of multiple control recipes in parallel. The authorised and configured master recipe procedure becomes the control recipe procedure when the batch has been assigned a batch number and has been approved for production. Control recipe execution may proceed in one of three operating modes: automatic, semi-automatic or manual.
- Resource management: The ability to control networked, multipath or single-path equipment configurations, to be able to manage complex batch

production facilities. The configuration of all items of equipment, shared-use equipment modules and exclusive-use equipment modules is integrated with the common object model. This makes adding and integrating a new process unit as simple as copying and pasting in the master control computer.
- Production scheduling and data collection: The batch management software provides a summary of all the batches in the current production queue. It can add batches to the real-time production schedule. It collects and records all data associated with and required for each batch.

Automated lubricant blending plants now use a DCS (Distributed Control System). This is a computerised control system with a number of control loops, in which autonomous controllers are distributed throughout the system, but there is central operator supervisory control. This is in contrast to non-DCSs that use centralised controllers: discrete controllers located either at a central control room or within a central computer. The DCS concept increases reliability and reduces installation costs by localising control functions near the process plant, with remote monitoring and supervision.

The key benefit of a DCS is its reliability, as a consequence of the distribution of the control elements throughout the system. If a single processor fails, it will only affect one section of the plant's process, as opposed to a failure of a central computer, which would affect the whole process. The distribution of computing power local to the individual input/output connections also enables fast control processing times, by removing possible network and central processing delays.

Most DCSs operate on five levels:

- Level 0: Field level, containing instrument (flow, temperature and other sensors) and final control elements, such as control valves and pumps.
- Level 1: Direct control level, containing input/output modules and their associated distributed electronic processors.
- Level 2: Plant supervisory level, containing the supervisory computers, which collect information from processor nodes on the system and provide the operator control screens.
- Level 3: Production control level, which does not directly control the process, but is concerned with monitoring production and targets.
- Level 4: Production scheduling level.

Levels 1 and 2 are the functional levels of a traditional DCS, in which all equipment is part of an integrated system from a single manufacturer. Levels 3 and 4 are not strictly process control in the traditional sense, but where production control and scheduling take place.

Signals from Level 0 are collected in a central location, sometimes called the motor control room (MCC), which is Level 1. Power to all the pump motors and control valves is controlled from the MCC. The control is achieved using PLCs, supplied by a number of companies, including Siemens, Emerson, Alan-Bradley and Yokogawa.

A DCS requires a communication network, typically a fieldbus. Initially, these used wires to connect all the system elements. More modern systems are now wireless. Fieldbuses have nearly always developed around a common theme of reduced wiring. However, the use of digital designs for field devices increased the benefits beyond wiring to enable maintenance practices and value-added asset management.

In a lubricant blending plant, the DCS can be configured as a single-user system or with client/server architectures, in which one batch server and several batch clients are used to control the operation of the blending plant. Software in the server and workstations provides operators with clear displays and diagrams which allow for the initiation and execution of blends during normal operation. Reports, alarms, messages, operator requests and acknowledgements are clearly displayed and logged, to provide a historical database of batches and to allow future analysis of health, safety, operational and customer service issues.

In the DCS, the PLCs interface with the computers that are in the client/server architecture. The field instruments, actuators and sensors are part of the fieldbus system. Electronic instrumentation, such as position switches and level, weight, pressure and temperature transmitters, are used as digital and analogue input to the PLCs and thus the DCS to provide data for alarms and control logic.

Supported technologies now include HART, WirelessHART, Modbus/Modbus TCP, PROFIBUS, PROFINET, FOUNDATION Fieldbus, Ethernet/IP, DeviceNet, Electrical Integration IEC 61850 and DNP3. Selecting the most appropriate fieldbus depends on the application and the system that is already installed. An optimised solution will probably use more than one fieldbus type. For example, new "analogue" installations benefit greatly from FOUNDATION Fieldbus, while new "discrete" installations can benefit from PROFIBUS, PROFINET, DeviceNet and Ethernet/IP. Electrical integration depends on the equipment-supported protocol, but IEC 61850, Ethernet/IP and PROFIBUS could all prove useful. The most common view is to look at the current installed base, whether HART, PROFIBUS, DeviceNet, Modbus or others. A lubricant blending plant may not need to completely replace a control system when it is possible to integrate new equipment with the existing equipment.

There are two main benefits to using a fieldbus. Many installations of fieldbuses have similar capital expenditure costs to conventional installations using remote inputs/outputs. Savings can be achieved with reduced installation and commissioning costs. Additionally, multifunction devices, for example, pressure–temperature–flow, can reduce the number of devices required and require only a single process controller. Lower operating costs are also achievable, by creating opportunities to implement maintenance strategies that optimise asset performance and utilisation, eliminate unnecessary maintenance, avoid unplanned interruptions to plant operations or extend plant turnaround cycles.

An illustration of a typical DCS in a lubricant blending plant is shown in Figure 7.18. This has a client/server architecture, fieldbus connections to different PLCs and standard interfaces to production planning systems. The client/server computers generally use the Windows operating system, now Windows 10, but previously Windows 8.1, 7, NT and XP. The programming languages for the PLCs and computers is generally according to standards ISA-S88 and IEC 1131 for the

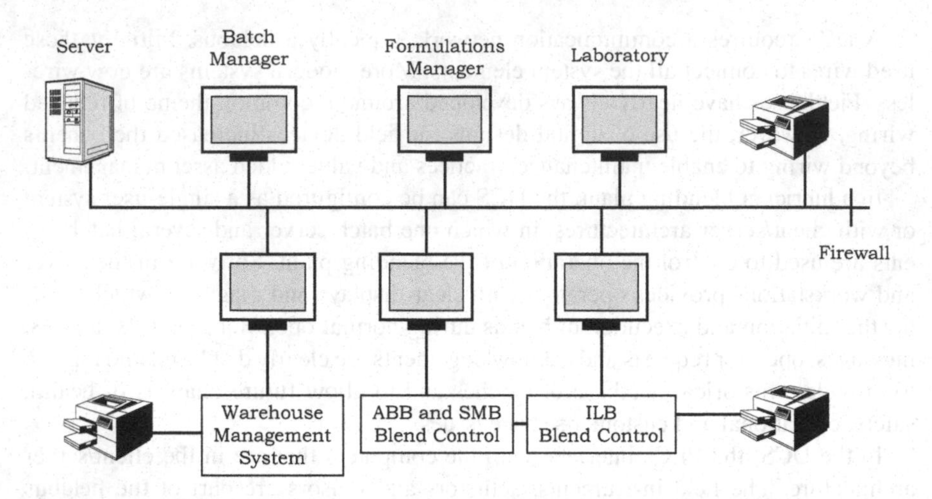

FIGURE 7.18 Lubricant blending plant distributive control system.

modulation of the batch structures, and the programmes are written by the suppliers of the PLCs.

The basic functions of the DCS include:

- Formulation selection and editing.
- Verification of tank size and quantity for blend quantity.
- Control of proper batch size and component additions.
- Automatic generation of batch number and batch ticket.
- Monitoring of all tank levels with appropriate alarms and interlocks.
- Automatic control of raw material and finished product pumps.
- Automatic control of raw material supply valves to the ABB and SMB.
- Prompting for manual additions of additives through the drum decanter.
- Automatic control of valves on ABB and SMB discharge headers.
- Automatic control of pigging operations.

One or more printers, connected to the DCS, generally in the MCC and the laboratory, are used for various printouts, including:

- Time and date for "Alarm on", "Alarm acknowledge" and "Alarm clear".
- Change of status (mode) of a loop tag, that is, auto to manual, manual to auto, local to remote, remote to local and so on.
- Change of set point or manual output.
- Any change in configuration parameters.
- Change of state, for example, start/stop.
- System diagnostics.
- Hard copy of system programme configuration.
- Batch tickets.
- Process termination, whether operator termination or due to a power failure.
- Reports: Batches, bulk inventory, packaging.

The DCS can be connected to the enterprise resource planning (ERP) system using Ethernet connections. Data can be transferred in both directions. This allows for complete integration between the DCS and the ERP system, which is discussed in more detail in Chapter 14.

The DCS enables the creation of individual production orders and batches of lubricants. A greatly increased planning functionality is offered by a batch planning option package. Batches can be scheduled in advance for a number of production orders generated by the ERP system.

In addition to planning, the scope of functions includes the modification, cancellation, deletion and enabling of batches. The creation and distribution of the batches for a production order are possible manually, but can also be carried out automatically depending on the definition of the batch number or production quantity.

Automation software can also include a formulation editor. This is a user-friendly tool for the easy modification of finished lubricant formulations and the creation of blending recipes and library operations. The basis for recipe creation is the batch objects created from the batch plant configuration, such as the ILB, SMB, ABB, DDU and pigging functions.

A batch report function integrated in the DCS will be used to produce formulation and batch reports. These can be displayed and printed. The batch reports contain all data required for reproduction of the lubricant batch process, for proof of quality and for compliance with statutory regulations.

7.7 SUMMARY

The combination of the available blending plant tools, such as ABBs, SMBs, ILBs and DDUs, adapts optimally to the production demands that are required by the market, as provided by optimum batch size calculations.

Modern lubricant blending plants are sophisticated and efficient, with state-of-the-art automated control systems. High levels of efficiency are necessary in order to fulfil the market's requirements. The key element is flexibility, to support the other company activities involved in providing high-quality products to customers, at the right time, in the right place and at the right price.

Optimised operation of a modern blending plant is a vital part of supply chain management. Careful attention to finished lubricant product quality, minimisation of the energy required to complete each blend, avoiding the need to reblend and minimising the generation of slop oil are very important for effective cost control.

A high degree of automation also means that modern blending plants can be operated with maximum efficiency, as well as maximum flexibility.

8 Lubricant Blending Issues
Avoiding Problems

8.1 INTRODUCTION

Blending lubricants may appear to be relatively straightforward. However, there are two major problems that need to be considered:

- Avoiding the need to reblend or correct an out-of-specification blend.
- Minimising to production of slop oil.

Additionally, there are several less important issues that also need to be avoided or minimised in order to operate a blending plant with optimum efficiency and profitability.

Actions to avoid or minimise problems with blending need to be taken before, during and after blending.

8.2 SAMPLING BEFORE BLENDING

Before blending starts, it is important to ensure that all the constituents in the formulation are as specified, as part of final product quality control.

To begin, all components must be sampled and analysed, preferably as soon as each batch is delivered to the blending plant. Various methods of sampling different items of equipment in a lubricant blending plant were presented and discussed in Chapter 7.

Three essential ways can be used to avoid problems with sampling base oils and additives prior to blending:

- Always use good-quality, clean, dry and correctly labelled sample bottles.
- Always use the same method to obtain a sample from a specific sampling point. Sampling procedures should be written as standard operating procedures (SOPs), to be used by all operators or supervisors when taking samples.
- Always obtain representative samples and never take a sample from an idle item of equipment.

A sample of each base oil or blend of base oils should be retained for at least 2 years. The results of the analytical tests on each batch of raw materials also need to be retained for at least 2 years, or preferably longer if they are stored electronically. Sample retention will assist in identifying any future customer complaint, since batch traceability (discussed in Chapter 11) will be easier if the properties and performance are known for all the raw materials that were used to blend the specific batch.

Additives delivered in bulk should be sampled in the same way as for base oils delivered in bulk. It is impractical to sample each drum or sack of an additive if multiple drums or sacks are delivered in one load.

Statistical sampling methods are appropriate. The number of drums or sacks from which samples are taken should be equal to the cube root of the total number of drums or sacks. For example, if eight drums are received, samples should be taken from two of them. Each batch of additive should be treated as a separate delivery, even if two or more batches arrive on the same truck.

Obviously, it is not generally practical or efficient for 100 drums or sacks of an additive to be delivered at one time. If a large quantity of an additive is required, it should be delivered in bulk, either in one or two intermediate bulk containers (IBCs), or road tankers, rail tank cars or barges, depending on the receiving facilities of the blending plant. Each of these bulk delivery methods should be treated as one batch for sampling and analysis, unless two or more of each delivery method arrive at the same time, in which case each should be sampled and tested independently, just in case there has been some contamination in one of the IBCs or tanks.

8.3 BLEND FAMILIES AND CROSS-CONTAMINATION

Some lubricants manufactured in blending plants have quite similar formulations, types of additives and performance requirements. The products comprise what have come to be called "blend families". Examples include:

- Automotive and industrial gasoline and diesel engine oils, both multigrade and monograde.
- Automotive and industrial gear oils, both monograde and multigrade
- Neat metalworking fluids.

These products can be blended using the same equipment without much risk that minor cross-contamination will degrade the performance properties of the finished lubricants.

Other lubricants have performance properties that are extremely susceptible to contamination by additives used in other lubricants. Example product types include:

- Hydraulic oils.
- Compressor oils.
- Turbine oils.
- Transformer and electrical oils.
- Water-mix metalworking fluids.

Each of these product groups should be blended in equipment that either can be dedicated to that group or can be cleaned of any residues from previous blending of other lubricant types.

The foaming and air-release properties of hydraulic oils are particularly susceptible to contamination by detergent and dispersant additives used in engine oils. The carbon residue properties of compressor oils may be degraded by the presence of viscosity index improving polymers. The corrosion resistance and metal passivation properties of turbine oils are susceptible to contamination by some anti-wear and extreme-pressure additives. Transformer and electrical oils must be kept very clean and dry and must not be contaminated by detergent or dispersant additives.

8.4 TEMPERATURE CONTROL

Controlling the temperatures of components before blending and of mixtures during and after blending is very important for ensuring an on-specification blend.

Many additives are susceptible to oxidation and thermal degradation, and so must not be heated too quickly or to temperatures above 60°C. However, some viscous viscosity index improvers may need to be heated to 80°C, but only for short periods (about 15 minutes) in order to lower their viscosity to the point where they are easier to blend.

All base oils and additives oxidise when exposed to air and heat. The Arrhenius equation indicates that the rate of oxidation doubles with every 10°C rise in temperature. Raw materials heated to 60°C will oxidise 4 times faster than those heated to 40°C; raw materials heated to 80°C will oxidise 16 times faster. In a blending plant, mixing at 60°C for an hour or so is unlikely to cause serious problems; mixing at 80°C for more than 15 minutes may result in significant oxidation of additives.

Many anti-wear and extreme-pressure additives thermally degrade at temperatures above 80°C. The surface temperatures of some electric heating coils can be as high as 120°C; at these temperatures, anti-wear and extreme-pressure additives will degrade rapidly. Low-pressure steam or hot-oil heating coils are much easier to control, particularly with regard to surface temperatures. It may take slightly longer to heat additives with hot oil or low-pressure steam compared with electricity, but off-specification products will be very much less.

8.5 TIMES FOR BLENDING

As noted in earlier chapters, the time taken to blend a specific lubricant or viscosity grade in an automatic batch blender (ABB) or simultaneous metering blender (SMB) will vary from the times taken for different lubricants or viscosity grades. For example, it is likely that an ISO 32 viscosity-grade HM (anti-wear) hydraulic oil can be blended in a shorter time than an ISO 100 viscosity-grade HM hydraulic oil, assuming that the blending temperatures are the same. At the same time, the time taken to blend an ISO 32 viscosity-grade HV (high viscosity index) hydraulic oil is likely to be longer than that for the ISO 32 viscosity-grade HM hydraulic oil, because the formulation for the former includes a viscosity index improver, which could take longer to dissolve in the blend.

The use of premix blending vessels will help to shorten blending times, particularly if a premix tank can be used for temporary (less than 24 hours) storage.

A premix blending vessel could be mixing a "cocktail" of additives for blending a product later in the day or shift, while an SMB is blending a different product, for example.

Blending plant managers and supervisors should maintain a database of blending times for every viscosity grade of every lubricant in every batch size (optimum blend size) that is blended in their facility. Some products may only take 30 minutes to blend in small quantities, while others may take 3 or 4 hours in large batch sizes. (Obviously, this does not apply to in-line blending [ILB].)

Knowing the time it takes to complete a specific blend, remembering to allow time for product transfer to holding tanks and for pigging operations, enables planning to schedule blending operations. For example, with an SMB that has four mixing vessels, it should be possible in one 8-hour shift to plan to complete four blends that take 1 hour each, another six blends that take 2 hours each and four more blends that take a total of 4 hours each. This equates to 14 blends in one shift. This helps to avoid wasted time and also helps the blending plant laboratory to plan for the tests that will be required for each blend. (The required tests are discussed in Chapter 10.)

8.6 MIXING

Several methods can be used to mix fluids. These include stirring (either slowly or quickly), high-shear mixing, air sparging or pumping around.

Stirring is the preferred method for mixing lubricating oils. This is usually done using a paddle mixer, as illustrated in Figure 7.1. The paddle mixer can have two, three or four straight blades, or the blades can be slightly curved, to introduce more "swirl" (turbulence) to the liquid being mixed. The paddle, usually placed near the bottom of the blending vessel, is rotated using an electrical motor, located at the top of the blending vessel. This is because the bottom of the blending vessel is conical in shape and has the outlet pipe at the lowest point of the sloping bottom, to minimise the generation of slop oil. The rotational speed of the paddle does not need to be particularly high.

Mixing a number of oils that have similar viscosities is relatively easy using a paddle mixer. Where there is a significant difference in viscosity between two liquids, a paddle mixer tends to move them around without achieving much mixing, and it can take a long time to achieve a uniform blend. High-shear mixers have a rotor–stator assembly that draws the liquids into the workhead where they are rapidly combined before being forced out through the stator, back into the vessel. The high-speed rotation of the rotor blades within a precision-machined mixing workhead exerts a powerful suction, drawing liquids upward from the bottom of the blending vessel and into the centre of the workhead. Centrifugal force then drives liquids toward the periphery of the workhead, where they are subjected to mixing in the clearance between the ends of the rotor blades and the inner wall of the stator. This is followed by hydraulic shear as the liquids are forced, at high velocity, out through the perforations in the stator and circulated into the main body of the mix. The liquids expelled from the workhead are projected radially toward the sides of the mixing vessel. At the same time, fresh material is continually drawn into the workhead, maintaining the mixing cycle. The horizontal (radial) expulsion and vertical suction sets up a circulation pattern that minimises aeration caused by the disturbance of the liquid's surface.

High-shear mixers are sometimes called "vortex mixers", although this term is more usually associated with small mixing units used in laboratories. These have a rapidly moving rubberised cup that helps to mix liquids in a test tube when it is held inside the cup.

High-shear mixers are also used in ILBs, for continuous homogenisation in a single pass, as described in Chapter 7. They can also be used to disperse powders into liquids. Different sizes of mixers are available, for use with different batch sizes or as part of premixing. Closed-rotor mixers are able to rapidly incorporate high percentages of powders and can handle relatively high-viscosity mixes. They are particularly suitable for incorporating difficult powders with a tendency to float or "raft".

However, high-shear mixers should be used with great care when mixing viscosity index improving polymers in multigrade lubricant formulations, particularly hydraulic oils and gear oils. The tendency of VI improvers to shear under high-shear stress means that high-shear mixers may need to be operated at relatively low speeds, to reduce the shear stress. Although this is likely to lengthen blending times, it will lessen the risk of the blend being off specification for viscosity and viscosity index properties.

Air sparging involves blowing compressed air through a perforated pipe, sparge ring or jet nozzle at the bottom of the blending vessel. A more sophisticated version of this is the "Pulsair" system, in which a programmed series of air bubbles provides efficient agitation. However, introducing air bubbles into a mixture of mineral or synthetic base oils and additives, particularly at temperatures of 60°C or above, risks oxidising the base oils and, more importantly, the additives. The susceptibility of some additives to oxidation is described in Chapter 4. Oxidation of the base oils and additives during air sparging may lead to the blend becoming off specification. Air sparging is used for blending fuels, such as gasoline or diesel, but as this is usually done at ambient temperatures, the risk of oxidation is low. Also, the fuels are likely to be used relatively soon following blending and distribution, whereas lubricants may still be lubricating items of equipment many years after blending and distribution.

Pumping around involves pumping the mixture from the bottom of the blending vessel and returning it to the blending vessel near the top of the vessel but below the level of the liquid in the vessel. This is so that the returning liquid does not fall into the liquid in the blending vessel and thereby introduce bubbles of air into the mixture. Again, introducing air bubbles into a blend of base oils and additives risks oxidation. Pump-around blending generally takes longer than paddle or vortex mixing, as the degree of turbulence in the mixture is lower.

A recent development with mixing fluids has been the introduction of "cold blending", in 2015, a method developed by a Russian engineering team associated with Polish GQOil Innovation Europe. The base oils and additives are first stirred together for 10–20 minutes in a premix tank and then fed into the blending unit's columns where the mixture is pumped at high velocity through a set of holes. As the moving liquid exits the holes under pressure, it releases air bubbles in a cavitation zone. These bubbles rapidly collapse due to the pressure of the surrounding oil, generating a burst of energy that disperses and homogenises the ingredients. The collapsing bubbles generate a jet of surrounding liquid and produce intense local heating, high pressures and then high cooling rates. It is claimed that the effect takes only a few microseconds so that the temperature rise from input to output is only between 2°C and 3°C. Any remaining bubbles move to the top of the liquid in a finishing tank, mechanically mixing the components in the process of de-aeration. The batch in the finished product tank is ready for testing, storage and packaging without any additional mixing.

A key advantage claimed for cavitation cold blending of lubricants (CCBL) is the reduced time and reduced temperature needed to accomplish mixing. Reduced blending time increases plant capacity and lower blending temperature reduces energy costs. CCBL allows blending to be performed at about 20°C, rather than the 40°C–60°C needed in conventional mixing processes.

At the time of writing, the process is being used by Prista Oil in its blending plant in Ruse, Bulgaria.[1] Tests on batches of engine oils manufactured using CCBL have demonstrated that "the technology and equipment secures smooth production and high blending efficiency", according to Prista Oil.

However, the author is aware that cavitation, particularly in hydraulic oils or steam turbine oils, can cause severe thermal and oxidative degradation and mechanical damage to servo valves and bearings. Surface temperatures on cavitating air bubbles can be as high as 600°C, even if only for milliseconds. It is possible that, unless great care is taken with the pressures, flow rates and blending times involved with CCBL, oxidatively and thermally sensitive additives, could be affected adversely. Also, depending on the shear rate through the holes in the blending unit's columns, viscosity index improving polymers may be sheared during blending. Further development and testing may be required for different types of lubricants before CCBL becomes more widely accepted in lubricant blending plants.

8.7 SAMPLING BLENDED LUBRICANTS

All blending equipment should have a sample point, preferably in the outlet line immediately before the pump-around line; samples should only be taken when the line is flowing. A sample of the blend should be taken when the final mixing appears to have been completed. If the blend is on specification, the product may be transferred to either a holding tank or product storage. If the blend is off specification, either the mixing has not been completed or incorrect dosing of raw materials has occurred.

Mixing should be continued, preferably with both the stirrer or the pump-around, for another 30 minutes and resampled. If the blend is now on specification, record the blending time required for that particular grade of product. If the blend is still off specification, the product will need to be reblended, after the blending plant laboratory has analysed the blend to determine which raw materials were dosed incorrectly.

8.8 SLOP OIL

In the oil industry, slop oil is the collective term for mixtures of oil, chemicals and water derived from a wide variety of sources in refineries or oil fields. It is formed when ships' tanks, oil tanks, pipelines or tank wagons are cleaned and during maintenance work or as a result of unforeseen oil accidents.

In a blending plant, slop oil is the material that results from the cleaning of blending tanks, storage tanks or pipelines. It is usually a mixture of base oils and blended lubricants.

Operators in many older blending plants use base oils to flush the pipelines between tanks, blending vessels and filling lines. These flushings, known as slop oil, can be used when blending some lubricants that do not have particularly high-performance properties. Unfortunately, with increasingly demanding performance requirements for many lubricants, using slop oil is becoming more difficult.

Generating slop oil can be greatly minimised by using pigging equipment. In addition, using dedicated blending equipment (for blend families) can reduce the need for flushing and cleaning significantly, which will thereby reduce the generation of slop oil. With in-line and simultaneous metering blending, in which base oils are used to flush transfer lines and these flushings are, in fact, part of the blend, there is no generation of slop oil.

With the increasing cost of base oils, generating slop oil is also an increasing waste of money. In a modern, efficient blending plant, the amount of slop oil that cannot be reused in products with lower-performance properties should be kept below 0.5% of the total volume of input raw materials. In other words, the total volume of on-specification products manufactured by a blending plant should be 99.5% of the volume of the input raw materials.

8.9 PACKAGES, LABELS AND LABELLING

All packages (bottles, cans and drums) of lubricants are required to be labelled by law, for compliance with health, safety and environmental regulations. Labels must either be stuck to the package, as with lubricant bottles; printed, as with tin-plate cans; or painted, as with drums.

Types of lubricant packages, the use of labels and methods for filling various packages are presented and discussed in Chapter 12. However, there are a number of issues that concern packages, labels and filling that are more appropriately covered in this chapter.

The outsides of packages must be kept clean and dry, so that the adhesive, printing ink or paint will adhere properly and stay there. Labels should be robust to the lubricant in the package. The ink, paint or adhesive should not degrade. Mislabelled packages or those whose labels have fallen off will need to be retested, at additional cost.

Damaged bottles, cans or drums that leak oil, or misapplied or unsealed closures (caps or bungs), will allow the build-up of a film of oil on the outside of the package. Although this is not a significant health, safety or environmental problem, it is likely to cause two other problems.

A film of oil on the outside of a package will attract dust and dirt, which may make it difficult to read the label. Oil, dust and dirt on the outside of a package is unsightly and may make it difficult to handle the package. Retail customers do not like to handle dirty and oily bottles. They begin to doubt the quality control and management of the company that manufactured the lubricant.

8.10 HEALTH, SAFETY AND THE ENVIRONMENT

As with all well-managed organisations, health, safety and the environment (HSE) should be important to all staff working in a lubricant blending plant. Managers and supervisors should ensure that all operations have minimal or no adverse effects on health, safety or the environment. A defined framework of methods and processes within the plant should guide all actions and embed a culture of personal and collective responsibility for HSE matters.

The written methods and processes should aim to ensure a safe and secure working environment, whereby all staff understand and follow the correct systems, procedures and conduct. The procedures need to comply fully with all international, national and regional laws or regulations that govern the handling of chemicals. In the case of a lubricant blending plant, this means base oils, additives and blended lubricants. In Europe, the relevant regulations are those of Registration, Evaluation and Authorisation of Chemicals (REACH). Companies manufacturing lubricants in the European Union (EU) or importing lubricants into the EU are required to comply with the REACH regulations. Comprehensive information, guidelines and contact details about REACH can be found on the European Chemicals Agency (ECHA) website: http://www.echa.europa.eu.

Most base oils and additives do not pose a health and safety risk. However, some need to be handled with special precautions, such as gloves and goggles. Suppliers of additives must always supply a safety data sheet (SDS) with each delivery of each additive. Blending plant managers, supervisors and operators need to be aware of all the precautions and handling instructions shown on the SDS.

Written health, safety and environmental procedures should be available to all staff in a lubricant blending plant and, if appropriate, placed in prominent positions in multiple locations, to explain what to do and who to contact in case of a spillage of base oils, additives or blended lubricants.

All bulk storage tanks for additives should have dedicated input and outlet pipelines. Pipelines for transferring base oils from tanks to blending vessels should be either dedicated or piggable. Pipelines for transferring additives from drum decanting units (DDUs) to blending vessels are cleaned with base oil, which forms part of the blend. Additives are comparatively expensive, so spillages are not only damaging to the environment but also costly.

8.11 FORKLIFT TRUCKS

The use of forklift trucks in storage warehouses is widespread and almost compulsory for manufacturing lubricants. Forklift trucks are used to move empty and full 205 L drums, empty and full IBCs and pallets of packed cartons and drums of finished oils. They are also used to deliver and remove drums and pallets from filling lines. Forklift trucks used in blending plants should have strong safety cabins; they should not be of the open type.

It is easier and safer for a forklift truck to move full drums that are stored on their sides, or to move pallets of drums, whether full or empty. Full 205 L drums, or pallets of drums, should be stored no more than two high, to lessen the risk of toppling a forklift truck. It is comparatively easy for a forklift truck, if not operated carefully, to puncture drums and cartons and damage conveyors used to move drums under filling heads. Filling line conveyors should have guard rails to prevent forklift trucks from getting too close; beware, however, that guard rails may present a hazard for filling line operators.

8.12 MINIMISING OPERATING EXPENSES

In a lubricant blending plant, operating expenses are generated by two principal factors:

- The manual involvement (supervision and operation) required in routine and automated processes.
- The inventory and production process.

People are required to be involved in the production process because the quality of information in the manufacturing execution system (MES) is not always sufficient or complete. They are also required because the blending programming or operating parameters are not routinely optimised.

Problems in the supply chain are evident from extended production times and increased inventories. Problems can result from:

- Unreliable suppliers.
- Out-of-specification raw materials.
- Insufficient or poor market information and forecasts.
- Incorrect programming or operating parameters.

Production time and inventory are only part of the dynamics. As a chain reaction, inventory means prolonged production time, and prolonged production time means increased inventory for basic materials and intermediate products, as well as finished products and work in process. It is quite challenging to measure production time, but inventory immediately translates into liquidity and cost. Therefore, it makes more sense to look at inventory as production time.

Unfortunately, minimising the operating expenses involved in lubricant blending is rarely straightforward. For example, having a low level of stock may conflict with optimising the size of production batches or maintaining a high level of customer service for variations in offtakes.

Each factor involved in production needs to be "weighted" against the company's strategic and tactical marketing and sales goals. For example, a company may decide that \$2 spent on storage costs is equivalent to \$1 spent on blending set-up costs. The aim of the complete optimisation should always be to minimise target costs.

Many factors, including such things as marketing promotions, new laboratory test procedures and new packaging designs' service level, may need to be considered in the solution of the conflict, according to their respective impacts.

Blending of lubricants can be completed more quickly using high-speed, high-shear mixers. Many types of lubricants can be blended very effectively using these mixers. However, great care needs to be used when using a high-shear mixer with any product that includes a viscosity index improving polymer, since there is a risk of shearing the polymer molecules. Lubricant types include multigrade engine oils, gear oils and hydraulic oils.

The time taken to complete a blend has a significant effect on the efficiency and cost-effectiveness of a blending plant. If a blend is mixed too quickly, it may not be completely homogeneous (and so be off specification) and the mixing time will have to be extended. If too much time is spent mixing a blend, energy will have been wasted (in both mixing and heating) and the blending equipment will not have been available for the next blend. Observations of blending times should be recorded for each grade of each product and an optimised blending time established.

Having to reblend should be avoided wherever possible. Reblending means using additional energy (a significant cost) and a potential reduction in the annual capacity of the blending plant. Reblending may also mean keeping a customer waiting for the delivery of a key product. Automated blending systems and effective quality control of raw materials should eliminate the need for reblending.

8.13 EQUIPMENT MAINTENANCE

One of the "unforeseen circumstances" that can cause significant problems for a lubricant blending plant is equipment breakdown or failure. Carefully planned maintenance of equipment can avoid breakdowns.

A blending plant, like any well-managed manufacturing facility, should have a comprehensive equipment monitoring and maintenance programme. The details of such a programme are beyond the scope of this book and would form the basis of a completely separate book.

Each item of equipment (ABBs, SMBs, ILBs, DDUs, pumps, valves and more) should be listed in a computerised database, with maintenance procedures, lubricant types (oils and greases), lubrication frequency, monitoring frequency, information reporting procedures and spare parts information and ordering. Planned equipment maintenance should be part of the process engineering operations of the blending plant.

An important part of planned maintenance should be the regular recalibration of load cells and mass flow meters. Several procedures for the calibration and recalibration of mass flow meters are described in Chapter 7. Similar procedures can be used for load cells. Manufacturers of load cells and mass flow meters, and the items of equipment that use them, will provide guidance on calibration methods, and some will provide recalibration services, either on site or off site. They will also advise on the optimum intervals between recalibration.

Of course, purchasing good-quality and reliable equipment is one way to guard against premature breakdown or failure of equipment. Most suppliers of the equipment used in a lubricant blending plant will provide information, guidance and assistance with equipment maintenance.

8.14 CYBER SECURITY

With the ever-increasing use of the Internet (World Wide Web) in all aspects of daily life, cyber attacks are being ranked as one of the most likely risks to global business, together with climate change and natural disasters. In an automated lubricant blending plant, the cyber security risks include theft of formulations, interference with blending operations, demands for payment from ransomware and even complete plant shutdown.

The lubricant plant's distributed control system (DCS), described in Chapter 7, is part of the information technology (IT) system that is almost certainly connected to the corporate (head office) IT system. The plant's warehouse management system (WMS), described in Chapter 13, is also part of the IT system. Malicious attacks on the corporate IT system may also result in attacks on the DCS and/or the WMS.

The aims of these systems are to improve efficiency, reduce costs, reduce downtime, achieve consistent product quality, avoid reblending, improve safety, meet regulatory requirements and protect brand reputation. These aims are at risk from both external and internal cyber attacks. Whether consciously or unconsciously, almost all automated processes have a direct or indirect connection to the Internet. The equipment ultimately connected to the Internet ranges from switches and pumps to programmable logic controllers, batch controllers, laboratory computers and printers and upward to the DCS and WMS.

Effective cyber security is not achieved by simply installing and relying on technology. Protection requires a combination of people, procedures and technology, and everyone in the blending plant must be involved. The main programmes required for industrial cyber security are listed in Table 8.1. A demilitarised zone (DMZ) on a router refers to a DMZ host, which is a host on the internal network that has all user datagram protocol (UDP) and transmission control protocol (TCP) ports open and exposed, except those ports otherwise forwarded. They are often used as a simple method to forward all ports to another firewall. In the field of computer security, security information and event management (SIEM) software products and services combine security information management (SIM) and security event management (SEM). They provide real-time analysis of security alerts generated by applications and network hardware. Continuous industrial control system (ICS) security

TABLE 8.1
Industrial Systems Cyber Security Programmes

Actions	Activities
Secure	Physical security, security practices, access control, asset inventory, device hardening, vulnerability management
Defend	DMZs, firewalls, unidirectional gateways, anti-malware, endpoint protection
Protect	Application whitelisting, zone firewalls, device firewalls
Monitor	SIEM, configuration change management, continuous ICS monitoring
Manage	Threat intelligence, incident management

Source: KEMSEC. With permission.

monitoring technologies provide defenders with the visibility needed. While some companies may not be aware of these solutions, others are already integrating them into their cyber security management programmes.

While the programmes listed in Table 8.1 are managed by a chief digital officer (an expanded role from the earlier chief information officer), all blending plant managers, supervisors and operators need to be aware of the threats and defences. The blending plant's executive directors should regard cyber security as a top priority in the current business environment.

With senior management support, the plant can adopt a security-centric approach. Cyber security is aligned with quality, safety, production, information management and all other functional activities. A framework for cyber security, published by the National Institute of Standards and Technology, is shown in Table 8.2.

The security-centric approach must also extend outside the blending plant. Automation of processes and communications are driving the creation and implementation of systems that involve suppliers and customers who are connected to what were once self-contained systems inside the plant. Cyber security needs to be applied across all these connections, as suppliers and customers will also want to be protected from malicious attacks. Conversely, the blending plant will want assurance from suppliers and customers about the robustness of their systems and processes across the supply chain reduces the risks and enables the creation of these extended systems. Enabling future digitisation and automation for plants and facilities is a priority across the all industries, as organisations seek to use information technology (IT) to become more efficient without affecting the safety and availability of their facilities.

Although many blending plants have invested in defence technologies, such as firewalls, it has been found that they have no visibility or protection for systems inside the firewall. Effective cyber security is achieved by a "defence-in-depth" approach that protects all components across the system by placing multiple layers of security. Perimeter defence alone leaves systems unprotected if this border is breached. It also leaves businesses unprotected in the event of an insider attack or in the event of an accidental threat. Accidental threats can be caused by non-malicious people, typically employees or contractors, who undertake actions that have the unintended

TABLE 8.2
National Institute of Standards and Technology Cyber Security Framework

Actions	Activities
Identify	Asset management, business environment, governance, risk assessment, risk management strategy
Protect	Awareness control, awareness and training, data security, information protection and procedures, protective technology
Detect	Anomalies and events, continuous security monitoring, detection processes
Respond	Responsive planning, communications, analysis, mitigation, improvements
Recovery	Recovery planning improvements, communications

Source: KEMSEC. With permission.

consequence of increasing risk. Effective cyber security enables the detection and prevention of these threats.

Suppliers of cyber security are able to use on-site surveys and discovery programmes to identify connections between systems and components. Once discovered, they can deploy solutions that provide protection across the DCS, WMS and IT system.

Further information on blending plant cyber security can be found in an informative article by Gareth Leggett.[2]

8.15 SUMMARY

Operating a blending plant and achieving cost-effective blending involves many more things than just mixing base oils and additives together.

There are a number of do's and don'ts to avoid or minimise problems:

Don't

- × Heat base oils or additives for too long or too hot before or during blending
- × Cross-contaminate products that have special or high-performance properties
- × Take unrepresentative samples of raw materials, blends or finished products

Do

- ✓ Check all raw materials for specified properties before blending
- ✓ Automate operations wherever possible
- ✓ Optimise the blending time for each grade of each product
- ✓ Check all products against defined properties and performance following blending
- ✓ Avoid having to correct a blend or, ultimately, reblend
- ✓ Avoid generating slop oil
- ✓ Practice good health, safety and environmental policies
- ✓ Maintain all items of equipment correctly, to avoid equipment breakdown or failure
- ✓ Implement an effective cyber security programme

Many ways exist to avoid or minimise problems that may occur before, during or after blending, packaging and filling.

REFERENCES

1. Kamchev, Boris, Prista Is Hot for Cold Blending, *Lube Report*, volume 17, issue 52, 9 September 2015, reprinted in *Prista Magazine*, issue 3, 2015, pp. 12–13.
2. Leggett, Gareth, Cyber Security in the Lubricants World, *Lubes 'n' Greases Europe-Middle East-Africa*, May 2018, pp. 11–14.

9 Testing and Analysis of Base Oils and Additives in Blending Plants

9.1 INTRODUCTION

All raw materials used in a blending plant to manufacture finished lubricants need to be tested before they are used, as part of the processes of quality control and quality assurance. The tests need not be exhaustive or time-consuming, and not all the tests described in this chapter are required for all base oils or additives.

A blending plant should aim to complete the selected tests as quickly as possible, and the tests should be selected based on their applicability and how long it takes to obtain a test result. Ideally, test times should be quite short and, if possible, completed while a base oil tank is being filled or a bulk additive tank is being filled. It is unlikely that a delivery driver will wait for test results on additives delivered in drums or sacks, but if the delivery driver is on a lunch break, this might be achievable.

Suppliers of raw materials should be asked to submit test results on their products as part of purchasing contracts, so only a few tests might be needed, simply to verify the accuracy of the suppliers' tests. Obviously, if any one of these initial tests reveals a raw material that is out of specification, a more extensive range of tests will be needed to ascertain why and whether corrective action can be considered or the raw material rejected.

Organisations that develop and publish standard test methods for use in the oil industry are listed in Table 9.1, together with their web page URLs. (They are also listed in the glossary at the end of this book.)

Standard test methods published by two of the organisations, the American Society for Testing and Materials (ASTM) and Institute of Petroleum (IP) (the Energy Institute [EI]) are the most commonly used in the oil industry. The ASTM develops test methods for a very wide range of industries, such as metals, plastics, water, petroleum, fibres, consumer products, microbiology, nanotechnology and manufacturing. More than 12,000 ASTM standards are used worldwide to improve product quality, enhance safety and facilitate trade. Individual standards, books that group similar standards together, several volumes covering an industry segment or the entire collection can be purchased, either in print or online. The ASTM has more than 140 technical standards-writing committees.

The EI was created in 2003 by the merger of two key organisations, the IP and the Institute of Energy (InstE). Both institutes had a distinguished heritage developed since the 1920s supporting their respective energy sectors. Increasingly, these sectors have converged, creating an integrated global energy market which has been

TABLE 9.1
Test Method Organisations

	Name	Website
ASTM	American Society for Testing and Materials	www.astm.org
IP	Institute of Petroleum (Energy Institute)	www.energyinst.org.uk
ISO	International Standards Organisation	www.iso.org
CEC	Coordinating European Council	www.cectests.org
CEN	Comité Européen de Normalisation	www.cen.eu
DIN	Deutsche Institut für Normung	www.din.de
ANSI	American National Standards Institute	www.ansi.org
AFNOR	Association Française de Normalisation	www.afnor.org
JSA	Japanese Standards Association	www.jsa.or.jp
JASO	Japanese Automotive Standards Organisation	www.jsae.or.jp

Source: Pathmaster Marketing Ltd.

mirrored by the development of the EI, which was established to address both the depth and breadth of the subject. The EI celebrated its centenary in 2014 to mark the creation of the oldest of the EI's founding organisations, the Institution of Petroleum Technologists. The EI is the leading chartered professional membership body for the energy industry, supporting more than 23,000 people working in or studying energy and 250 companies worldwide.

IP test methods were developed over many years by the IP. IP test methods have become so well established in the oil industry that the EI decided to retain the nomenclature IP for the tests. Many of the ASTM and IP test methods for petroleum products (fuels, lubricants, gases, bitumens and waxes) and crude oils are either identical or are published as joint standards. The two organisations have joint working groups and committees that develop the test methods.

9.2 TESTS FOR BASE OILS

Tests on base oil samples should follow those given in specifications, but not every test may need to be done prior to using the base oil in a blending plant. Only those tests that are most usually done to determine the properties of a specific shipment of a base oil are described. The details of the specific test methods can be found in the relevant books and manuals published by the associations listed in Table 9.1. The designations of the most commonly used test methods are also shown below.

Three properties of Group I, II and III base oils are assumed as standard requirements:

- Clear and bright in appearance, that is, not cloudy.
- A maximum water content of 100 parts per million (ppm).
- A maximum acid number (AN) of 0.05 mg of potassium hydroxide (KOH) per g.

The first two of these "standard" properties apply also to the synthetic base oils described in Chapter 3. If a base oil shipment does not meet these requirements, it must not be used to blend lubricants and must not be mixed with the same type and viscosity of a base oil in storage in the blending plant. The reason for its non-conformance needs to be determined, almost certainly in discussion with the supplier.

9.2.1 Colour

The colours of base oils, as observed by light transmitted through them, vary from practically clear or transparent to opaque or very dark brown. The methods of measuring colour are based on a visual comparison of the amount of light transmitted through a specified depth of oil with the amount of light transmitted through one of a series of coloured glasses. The colour is then given as a number corresponding to the number of the coloured glass. The two most commonly used methods for determining the colour of a base oil are ASTM D1500 and IP 196, which are technically equivalent.

Colour variations in base oils result from differences in crude oils, viscosities and methods and degrees of treatment during refining. During processing, colour is a useful guide to a refiner to indicate whether processes are operating properly. In finished lubricants, colour has little significance except in the case of medicinal and industrial white oils, which are often compounded into, or applied to, products in which staining or discoloration would be undesirable.

9.2.2 Boiling Range

The boiling range of a base oil is a particularly good way of establishing whether it has been fractionated (distilled) properly during manufacturing. Previously, the distillation range was determined using distillation in laboratory apparatus. The current method, ASTM D2887, uses gas chromatography.

The sample is introduced into a gas chromatographic column, which separates hydrocarbons in boiling point order. The column temperature is raised at a reproducible rate, and the area under the chromatogram is recorded throughout the run. Boiling temperatures are assigned to the time axis from a calibration curve, obtained under the same conditions by running a known mixture of hydrocarbons covering the boiling range expected in the sample. From these data the boiling range distribution may be obtained.

The standard method (ASM D2887) covers determination of the boiling range distribution of petroleum products and fractions with a final boiling point (FBP) of 538°C or lower at atmospheric pressure. The method is limited to samples having a boiling range greater than 55°C and having a vapour pressure sufficiently low to permit sampling at ambient temperature.

The initial boiling point (IBP) is the temperature at which a cumulative area count equal to 0.5% of the total area under the chromatogram is obtained. The FBP is the temperature at which a cumulative area count equal to 99.5% of the total area under the chromatogram is obtained.

9.2.3 Density

The density of a substance is the mass of a unit volume of it at a standard temperature. The specific gravity (relative density) is the ratio of the mass of a given volume of a material at a standard temperature to the mass of an equal volume of water at the same temperature. American Petroleum Institute (API) gravity is a special function of specific gravity (SG), which is related to it by the following equation:

$$\text{API gravity} = \frac{141.5}{\text{SG } 60/60°\text{F}} - 131.5$$

The API gravity value therefore increases as the specific gravity decreases. Since both density and gravity change with temperature, determinations are made at a controlled temperature and then corrected to a standard temperature by use of special tables.

Density determinations are made quickly and easily. There are three standard methods: IP 190 (capillary-stoppered pyknometer), ASTM D1217/IP 249 (Bingham pyknometer) and ASTM D4052/IP 365 (digital density meter). The last method is most useful for a lubricant blending plant because it is rapid and automated. The standard temperature for petroleum products, including base oils, is 15°C. Density is useful for identifying oils, provided the distillation range or viscosity of the oils is known. The primary use of API gravity, however, is to convert weighed quantities to volume and measured volumes to weight.

9.2.4 Kinematic Viscosity

Probably the most important single property of a lubricating oil is its viscosity. A factor in the formation of lubricating films under both thick- and thin-film conditions, viscosity affects heat generation in bearings, cylinders and gears. It also determines the ease with which machines may be started under cold conditions, and it governs the sealing effect of the oil and the rate of consumption or loss. For any piece of equipment, the first essential for satisfactory results is to use an oil of proper viscosity to meet the operating conditions.

The basic concept of viscosity can be illustrated by moving a plate at a uniform speed over a film of oil. The oil adheres to both the moving surface and the stationary surface. Oil in contact with the moving surface travels with the same velocity (U) as that surface, while oil in contact with the stationary surface has zero velocity. In between, the oil film may be visualised as being made up of many layers, each drawn by the layer above it at a fraction of velocity U that is proportional to its distance above the stationary plate. A force F must be applied to the moving plate to overcome the friction between the fluid layers. Since this friction is the result of viscosity, the force is proportional to viscosity. Viscosity can be determined by measuring the force required to overcome fluid friction in a film of known dimensions. Viscosity determined in this way is called dynamic or absolute viscosity.

Dynamic viscosities are usually reported in poise (P) or centipoise (cP) (1 cP = 0.01 P), or in SI units in pascal-seconds (Pa-s) (1 Pa-s = 10 P). Dynamic viscosity, which is a function only of the internal friction of a fluid, is the quantity used most frequently in bearing design and oil flow calculations. Because it is more convenient to measure viscosity in a manner such that the measurement is affected by the density of the oil, kinematic viscosities normally are used to characterise lubricants.

The kinematic viscosity of a fluid is the quotient of its dynamic viscosity divided by its density, both measured at the same temperature and in consistent units. The most common units for reporting kinematic viscosities are stokes (St) or centistokes (cSt) (1 cSt=0.01 St), or in SI units, square millimetres per second (mm^2/s) (1 mm^2/s=1 cSt). Dynamic viscosities, in centipoise, can be converted to kinematic viscosities, in centistokes, by dividing by the density in grams per cubic centimetre (g/cm^3) at the same temperature. Kinematic viscosities, in centistokes, can be converted to dynamic viscosities, in centipoise, by multiplying by the density in grams per cubic centimetre. Kinematic viscosities, in square millimetres per second (mm^2/s), can be converted to dynamic viscosities, in pascal-seconds, by multiplying by the density, in grams per cubic centimetre, and dividing the result by 1000.

Other viscosity systems, including those of Saybolt, Redwood and Engler, are in use and will probably continue to serve for many years because of their familiarity to some people. However, the instruments developed to measure viscosities in these systems are rarely used. Most actual viscosity determinations are made in centistokes and converted to values in the other systems by means of published SI conversion tables.

The viscosity of any fluid changes with temperature, decreasing as the temperature is increased and increasing as the temperature is decreased. It is therefore necessary to have some method of determining the viscosities of lubricating oils at temperatures other than those at which they are measured. This is usually accomplished by measuring the viscosity at two temperatures and then plotting these points on special viscosity–temperature charts developed by the ASTM. A straight line can then be drawn through the points and viscosities at other temperatures read from it with reasonable accuracy. The line should not be extended below the pour point or above approximately 150°C (for most lubricating oils) since in these regions it may no longer be straight.

The standard methods for measuring kinematic viscosity are ASTM D445 and IP 71, which are technically equivalent. The two temperatures most used for measuring and reporting kinematic viscosities are 40°C (104°F) and 100°C (212°F). Dynamic viscosities are measured at several different temperatures; for base oils and automotive engine oils, these are usually in the range –20°C–-35°C. Lubricant blending plants do not usually need to measure the dynamic viscosity of a base oil.

9.2.5 Viscosity Index

Different oils have different rates of change of viscosity with temperature. For example, a distillate oil from a naphthenic crude oil would show a greater rate of change of viscosity with temperature than would a distillate oil from a paraffinic crude oil. The viscosity index (VI) is a method of applying a numerical value to this rate of change, based on comparison with the relative rates of change of two arbitrarily selected types of oil that differ widely in this characteristic. A high VI indicates a relatively low rate of change of viscosity with temperature; a low VI indicates a relatively high

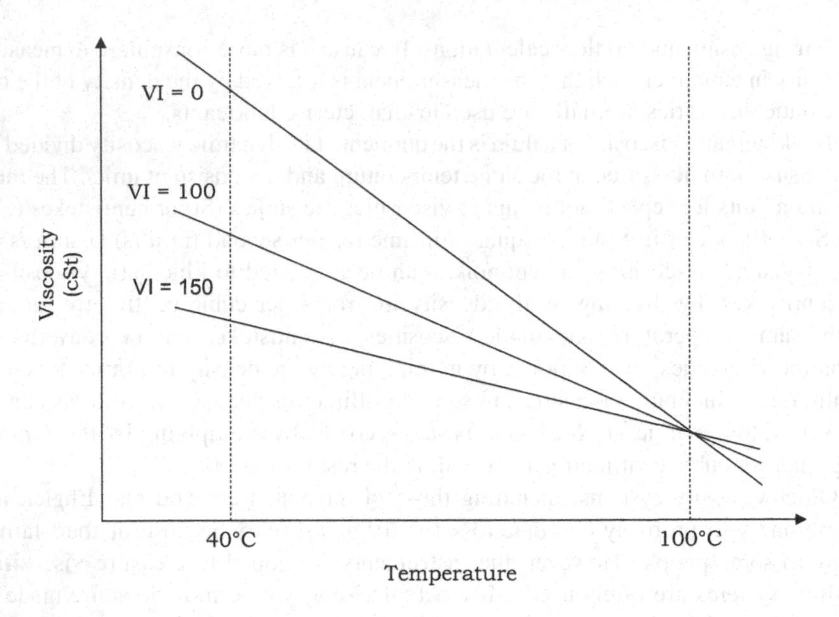

FIGURE 9.1 Graphical representation of viscosity index. (Source: Pathmaster Marketing Ltd.)

rate of change of viscosity with temperature. For example, consider a high-VI oil and a low-VI oil having the same viscosity at, say, room temperature: as the temperature increased, the high VI oil would thin out less and therefore would have a higher viscosity than the low-VI oil at higher temperatures.

The VI of an oil is calculated from kinematic viscosities determined at two temperatures by means of tables published by the ASTM (ASTM D2270) and the EI (IP 226). Tables based on kinematic viscosities determined at both 40°C and 100°C and 100°F and 212°F are available. A diagrammatic representation of VI is shown in Figure 9.1.

Mineral oil base oils refined through special hydroprocessing techniques can have VIs well above 100. Some synthetic lubricating oils have VIs both below and above this range, as discussed in Chapters 2 and 3.

9.2.6 Low-Temperature Viscosity

Rotational viscometers measure dynamic viscosity, using the viscous drag on a rotor immersed in oil and measuring the torque required to rotate it at a given speed, or the speed achieved for a given torque. Important examples for base oils and automotive engine oils include the Brookfield viscometer, the cold cranking simulator (CCS) and the mini-rotary viscometer (MRV), all for low-temperature use. (The tapered bearing simulator [TBS] is used for measuring high-shear dynamic viscosity at 150°C, applicable only to engine oils.)

The Brookfield viscometer, test method ASTM D2983, is a rapid, direct-reading instrument employing an assortment of rotary spindles and variable-speed settings (very low to medium) to provide a wide range of measurements. The viscometer has a spindle driven by a motor which is able to measure the torque required to turn the spindle. The test oil is cooled in an air bath for 16 hours at the test temperature and taken to the viscometer. The spindle is immersed in the test liquid and the speed of rotation selected, and the torque reading is a measure of the apparent (dynamic) viscosity.

Between 1957 and 1967, a viscometer with a very specific use was developed. This was the CCS, whose purpose was to define the winter grades of motor oil in a way that correlated with engine crankability, rather than by use of the VI system. A universal motor drives a rotor which is closely fitted inside a stator; a small sample of oil fills the space between the rotor and stator. The speed of the rotor is a function of the oil's viscosity. From a calibration curve and the measured speed of the rotor, the viscosity of the test oil, in centipoise, is calculated. The CCS was originally operated at a single temperature, namely 0°F or –18°C.

Recent modifications to the viscosity classification system now require test temperatures ranging from –5°C for 25 W oils to –30°C for 0 W oils. The method is given in ASTM D5293 and IP 350, which are technically equivalent. A particular type of electric motor applies a relatively constant torque to the rotor, and the speed of rotation is related to the viscosity by calibration with standard oils.

In the 1980s, problems of oil pumpability in cases where cranking was satisfactory led to further modifications of the viscosity classification, and the adoption of new low-temperature viscometers. Low-temperature, low-shear viscosity is important for predicting the possibility of "air binding" in motor oils after vehicles have stood at low temperatures for a considerable period. The non-Newtonian motor oil can gel to a semi-solid and fail to flow to the oil pump inlet when the engine is started. The oil pump then pumps air instead of oil to the engine, and both the pump and other engine parts can be rapidly damaged. Even if "air binding" does not take place, an oil can be so viscous after standing at low temperatures that the rate of pumping oil to sensitive bearings and rockers may be inadequate, and again engine damage can result. The Brookfield method ASTM D5133 is believed to correlate with these problems, and it is recommended that this test is performed on new oil formulations. It is, however, time-consuming and does not readily permit tests on large numbers of samples, and so is not applicable for use in lubricant blending plants.

For base oils, low-temperature flow properties are a better guide as to their suitability for use in automotive engine oils, automatic transmission fluids and some gear oils and hydraulic oils than is pour point.

9.2.7 Pour Point

The pour point of a lubricating oil is the lowest temperature at which it will pour or flow when it is chilled, without disturbance, under prescribed conditions. Most mineral base oils contain some dissolved wax, and as an oil is chilled, this wax begins to separate as crystals that interlock to form a rigid structure that traps the oil in small pockets in the structure. When this wax crystal structure becomes sufficiently complete, the oil will no longer flow under the conditions of the test. Since, however, mechanical agitation can break up the wax structure, it is possible to have an oil flow at temperatures considerably below its pour point. Cooling rates also affect wax crystallisation; it is possible to cool an oil rapidly to a temperature below its pour point and still have it flow.

Three test methods can be used to determine the pour point of a base oil. In methods IP 15 and ASTM D97, a sample of oil is put into a small test tube and, after preliminary heating, the sample is cooled at a specified rate and examined manually at intervals of 3°C for flow characteristics. The pour point of the oil is the temperature 3°C above the temperature at which the oil no longer flows. In an automatic pressure pulsing method (ASTM D5949), the test oil is put into the automatic test apparatus and is heated and then cooled by a Peltier device at a rate of 1.5°C ± 0.1°C. At temperature intervals of 1°C, 2°C or 3°C (selected by the technician), a pressurised pulse of nitrogen gas is imparted to the surface of the oil. Multiple optical detectors are used in conjunction with a light source to monitor movement of the oil's surface. The lowest temperature at which movement of the oil's surface is detected is recorded as the pour point. Method ASTM D5895 is a rotational test in which the test oil is put into the automatic test apparatus and is heated and then cooled by maintaining a constant temperature differential between a cooling block and the test oil. The oil is continuously tested for flow characteristics by rotating the test cup at about 0.1 rpm against a stationary, counterbalanced, sphere-shaped pendulum. The temperature of the oil at which a crystal structure or viscosity increase causes a displacement of the pendulum is recorded as the pour point with a resolution of 0.1°C.

Again, the last two test methods are of more value in a lubricant blending plant than the first test method, because they can be run more quickly, although all three methods are now automated.

While the pour point of most oils is related to the crystallisation of wax, certain oils, which are essentially wax-free, have viscosity-limited pour points. In these oils, the viscosity becomes progressively higher as the temperature is lowered until at some temperature no flow can be observed. The pour points of such oils cannot be lowered with pour point depressants, since these agents act by interfering with the growth and interlocking of the wax crystal structure.

Base oils show wide variations in pour points. Distillates from waxy paraffinic or mixed base crude oils typically have pour points in the range of 27°C–49°C, while raw distillates from naphthenic crude oils may have pour points on the order of −18°C or lower. After solvent dewaxing, the paraffin distillates will have pour points on the order of −7°C–−18°C. Where lower pour points oils are required, pour point depressants are normally used.

9.2.8 Flash Point

The flash point of an oil is the temperature at which the oil releases enough vapour at its surface to ignite when an open flame is applied. In methods IP 36 and ASTM D92 (the Cleveland open cup [COC] test), a test cup is filled with the sample to a specified level. The temperature is increased rapidly at first and then at a slow constant rate as the flash point is approached. At specified intervals, a small test flame is passed across the cup. When the concentration of vapours at the surface becomes great enough, exposure to an open flame will result in a brief flash as the vapours ignite. The temperature at which this happens is the flash point.

Methods IP 34 and ASTM D93 (the Pensky–Martens closed cup [PMC] test) heat the sample of oil at a slow, constant rate with continual stirring in a special "closed cup". A small flame is directed into the cup at regular intervals with simultaneous interruption of stirring. Again, the flash point is the lowest temperature at which application of the flame causes the vapour above the sample to ignite.

The differences between the two methods are significant. In the PMC test, any low-volatility contaminants in a base oil will accumulate more quickly in the closed test cup, resulting in a lower flash point than would otherwise have been the case. The PMC method can therefore be used to determine if a base oil has become contaminated prior to its receipt in a lubricant blending plant. The COC test is used more often to determine an inherent property of the base oil, and so is of less use in a lubricant blending plant.

These methods should be used to measure the properties of oils in response to heat and flame, but not to appraise the fire hazard of oils under fire conditions. The release of vapours at the flash point is not sufficiently rapid to sustain combustion, and so the flame immediately dies out. However, if heating is continued, a temperature will be reached at which vapours are released rapidly enough to support combustion. This temperature is called the fire point. Method IP 35 can be used to determine the fire point of petroleum products.

The flash point of new oils varies with viscosity; higher-viscosity oils have high flash points. Flash points are also affected by the type of crude and by the refining process. For example, naphthenic oils generally have lower flash points than paraffinic oils of similar viscosity. Flash point tests are of value to refiners for control purposes and are significant to consumers under certain circumstances for safety considerations.

9.2.9 Volatility

There is a direct association with an oil's volatility characteristics and oil consumption rates. Although volatility is not the sole reason for oil consumption in any given engine, it provides a measure of the oil's ability to resist vaporisation at high temperatures. Typically, distillations were run to determine volatility characteristics of base stocks used to formulate engine oils. This is still true currently. The objective of further defining an oil's volatility led to the introduction of several new non-engine bench tests. The most common of these tests are the NOACK volatility, simulated distillation, or "sim-dis" (ASTM D2887), and the GCD (gas chromatography distillation) volatility (ASTM D5480) tests. All these tests measure the amount of oil in percent that is lost upon exposure to high temperatures and therefore serve as a measure of the oil's relative potential for increased or decreased oil consumption during severe service.

Reduced volatility limits are placing more restraints on base oil processing and selection and the additive levels to achieve the required volatility levels. Volatility is one of the properties of a base oil that cannot be corrected by the use of additives, so having a low-volatility base oil as the basis for a low-volatility automotive engine oil is critically important.

9.2.10 Foaming Properties

In the most widely used foam test (ASTM D892), air is blown through an oil sample held at a specified temperature for a specified period of time. Immediately after blowing is stopped, the amount of foam is measured and reported as the foaming tendency. After a specified period of time to allow for the foam to collapse, the volume of foam is again measured and reported as the foam stability.

This test gives a fairly good indication of the foaming characteristics of new, uncontaminated oils. Test results may not correlate well if contaminants such as moisture or finely divided rust, which can aggravate foaming problems, are present in a system. It is important for a lubricant blending plant's managers, supervisors and operators to understand that base oils may be delivered with some water and/or rust contamination, which may not be apparent when examined visually. A test of the foaming properties of newly delivered base oils may uncover any contamination.

For a number of lubricant applications, such as automatic transmission fluids, special foam tests involving severe agitation of the oil have been developed. These are generally proprietary tests used for specification purposes.

9.2.11 Demulsibility

Lubricating oils used in circulating systems should separate readily from water that may enter the system as a result of condensation, leakage or splash of process water. If the water separates easily, it will settle to the bottom of the reservoir, where it can be periodically drawn off. Steam turbine oils, hydraulic fluids and industrial gear oils are examples of products for which it is particularly important to have good water separation properties. These properties are usually described as the demulsibility or demulsification characteristics.

The ASTM D2711, "Demulsibility Characteristics of Lubricating Oils", is intended for use in testing medium- and high-viscosity oils, such as paper machine oil applications, that are prone to water contamination and may encounter the turbulence of pumping and circulation capable of producing water-in-oil emulsions. A modification of the test is suitable for evaluation of oils containing extreme-pressure additives. In the test, water and the oil under test are stirred together for 5 minutes at 82°C. After the mixture has been allowed to settle for 5 hours, the percentage of water in the oil and the percentages of water and emulsion separated for the oil are measured and recorded.

IP 19, "Demulsification Number", is intended for testing new inhibited and straight mineral turbine oils. A sample of the oil held at about 90°C is emulsified with dry steam. The emulsion is then held in a bath at 94°C and the time in seconds for the oil to separate is recorded and reported as the demulsification number. If the complete separation has not occurred in 20 minutes, the demulsification number is reported as 1200+.

For base oils, a low value for demulsibility or demulsification number indicates that the oil has not become contaminated during refining, storage or transportation.

9.2.12 Acid Number

The AN of an oil is synonymous with neutralisation number. The AN of an oil is the weight in milligrams of KOH required to neutralise 1 g of oil and is a measure of all the materials in an oil that will react with KOH under specified test conditions. The usual major components of such materials are organic acids, soaps of heavy metals, intermediate and advanced oxidation products, organic nitrates, nitro compounds and other compounds that may be present as additives.

Since a variety of degradation products contribute to the AN value, and since the organic acids present vary widely in corrosive properties, the test cannot be used to predict corrosiveness of an oil under service conditions.

Tests were developed to provide a quick determination of the amount of acid in an oil by neutralising it with a base. The amount of acid in the oil was expressed in terms of the amount of a standard base required to neutralise a specified volume of oil. This quantity of base came to be called the neutralisation number of the oil.

The two most commonly used tests for acidity (AN) are ASTM D664 (equivalent to IP 177) and ASTM D974 (equivalent to IP 139). Both test methods can be used to measure total AN, strong AN, total base number and/or strong base number.

9.2.13 Carbon Residue

The carbon residue of an oil is the amount of deposit, in percentage by weight (% wt), left after evaporation and pyrolysis of the oil under prescribed conditions. In the test, oils from any given type of crude oil show lower values than those of similar viscosity containing residual stocks. Oils of naphthenic type usually show lower residues than those of similar viscosity made from paraffinic crude oils. The more severe the refining treatment (whether an oil is subjected to solvent processing or hydroprocessing), the lower the carbon residue value will be.

Originally, the carbon residue test was developed to determine the carbon-forming tendency of steam cylinder oils. Subsequently, unsuccessful attempts were made to relate carbon residue values to the amount of carbon formed in the combustion chambers and on the pistons of internal combustion engines. Since such factors as fuel composition and engine operation and mechanical conditions, as well as other lubricating oil properties, are of equal or greater importance, carbon residue values alone have only limited significance. The carbon residue determination is now made mainly on base oils used for engine oil manufacture, straight mineral engine oils such as aircraft engine oils, and some products of the cylinder oil type used for reciprocating air compressors. In these cases, the determination is an indication of the degree of refining to which the base oil has been subjected.

ASTM D4530, "Standard Test Method for Determination of Carbon Residue (Micro Method)", measures the "Conradson carbon residue" of a base oil or lubricant. The method offers advantages of better control of test conditions, smaller samples and less operator attention compared with test method ASTM D189, to which it is equivalent.

The test covers the determination of the amount of carbon residue formed after evaporation and pyrolysis of petroleum materials under certain conditions and is intended to provide some indication of the relative coke-forming tendency of such materials. Up to 12 samples may be run simultaneously, including a control sample when a vial holder shown in the test method is used exclusively for sample analysis.

The procedure is a modification of the original method and apparatus for carbon residue of petroleum materials, where it has been demonstrated that thermogravimetry is another applicable technique. However, it is the responsibility of the operator to establish operating conditions to obtain equivalent results when using thermogravimetry.

The test is applicable to petroleum products that partially decompose on distillation at atmospheric pressure and was tested for carbon residue values of 0.10% wt to 30% wt. Samples expected to be below 0.10% wt residue should be distilled to remove 90% v/v of the flask charge. The 10% bottoms remaining are then tested for carbon residue.

Ash-forming constituents, as defined by test method D482 (see later), or non-volatile additives present in the sample will add to the carbon residue value and be included as part of the total carbon residue value reported. The test results are equivalent to ASTM D189 (IP 13).

ASTM D524, "Standard Test Method for Ramsbottom Carbon Residue of Petroleum Products", is another method for determining the carbon residue of base oils and lubricants. The carbon residue value of motor oil, while at one time regarded as indicative of the amount of carbonaceous deposits a motor oil would form in the combustion chamber of an engine, is now considered to be of doubtful significance due to the presence of additives in most oils. For example, an ash-forming detergent additive can increase the carbon residue value of an oil yet will generally reduce its tendency to form deposits.

The carbon residue values of crude oil atmospheric residues, cylinder oils and bright stocks are useful in the manufacture of lubricants.

The test also covers the determination of the amount of carbon residue left after evaporation and pyrolysis of an oil, and it is intended to provide some indication of relative coke-forming propensity. The test is generally applicable to relatively non-volatile petroleum products which partially decompose on distillation at atmospheric pressure. This test method also covers the determination of carbon residue on 10% v/v distillation residues. As with ASTM D4530, petroleum products containing ash-forming constituents as determined by ASTM D482 will have an erroneously high carbon residue, depending on the amount of ash formed.

The term *carbon residue* is used throughout the test to designate the carbonaceous residue formed during evaporation and pyrolysis of a petroleum product. The residue is not composed entirely of carbon but is a coke which can be further changed by pyrolysis. The term *carbon residue* is continued in this test method only in deference to its wide common usage.

The values obtained by the test are not numerically the same as those obtained by ASTM D189 or ASTM D4530. Approximate correlations have been derived but need not apply to all materials which can be tested because the carbon residue test is applicable to a wide variety of petroleum products. The Ramsbottom carbon residue test method is limited to those samples that are mobile below 90°C.

9.2.14 Water Content

A simple way to check for the presence of water droplets in a base oil is to use the "crackle test". The end of a heated rod is dipped into the oil and if the oil is heard to "crackle", free water is present. (This can often be determined simply by the hazy appearance of the oil.)

Two, more accurate methods are used to determine the water content of base oils or lubricants. Method ASTM D6304, Karl Fischer titration, is used to determine the water content of base oils and petroleum products by coulometric titration. The method uses standard automatic Karl Fischer titration equipment and reagents which allow the determination of water contents in the range of 10–25,000 mg/kg (ppm) entrained water. The method also covers the indirect analysis of water thermally removed from samples and swept with dry inert gas into the Karl Fischer titration cell. The presence of sulphur, sulphides (including H_2S) and mercaptans is known to interfere with the result.

An alternative is method IP 358 (ASTM D4006), distillation. The method was originally developed for determining water in crude oil, but it may be applied to other petroleum hydrocarbons. The sample is heated under reflux with a water-immiscible solvent which co-distils with the water in the sample. Condensed solvent and water are continuously separated in a trap, the water settling in the graduated section and the solvent returning to the still. The method is similar to IP 74 (ASTM D95).

9.2.15 Sulphur, Nitrogen and Phosphorous Contents

ASTM D5291, "Standard Test Methods for Instrumental Determination of Carbon, Hydrogen, and Nitrogen in Petroleum Products and Lubricants", is the first ASTM standard covering the simultaneous determination of carbon, hydrogen and nitrogen in petroleum products and lubricants. Carbon, hydrogen and particularly nitrogen analyses are useful in determining the complex nature of petroleum products covered by the test.

The concentration of nitrogen is a measure of the presence of nitrogen-containing additives. Knowledge of its concentration can be used to predict performance. Some petroleum products also contain naturally occurring nitrogen. Knowledge of hydrogen content in samples is helpful in addressing their performance characteristics. The hydrogen-to-carbon ratio is useful to assess the performance of upgrading processes.

The test method covers the instrumental determination of carbon, hydrogen and nitrogen in laboratory samples of petroleum products and lubricants. Values obtained represent the total carbon, the total hydrogen and the total nitrogen. The test was tested in the concentration range of at least 75–87 mass percent for carbon, at least 9–16 mass percent for hydrogen and < 0.1–2 mass percent for nitrogen.

The nitrogen test method is not applicable to light materials or those containing < 0.75 mass percent nitrogen, or both, such as gasoline, jet fuel, naphtha, diesel fuel or chemical solvents. However, using ASTM D5291 levels of 0.1 mass percent nitrogen in lubricants could be determined. The test method is not recommended for the analysis of volatile materials such as gasoline, gasoline–oxygenate blends or gasoline-type aviation turbine fuels.

The test instrument works on the basis of a combustion method to convert the sample elements to simple gases (CO_2, H_2O and N_2), at an elevated temperature (975°C) and in a pure oxygen environment.

Elements, such as halogens and sulphur, are removed by scrubbing reagents in the combustion zone. The combustion product gas stream is passed over a heated reduction zone, which contains copper to remove excess oxygen and reduce NOx to N_2 gas. The gases are then homogenised and controlled, in a mixing chamber, to exact conditions of pressure, temperature and volume. The homogenised gases are allowed to de-pressurise through a column where they are separated in a stepwise steady-state manner and detected as a function of their conductivities.

ASTM D4294, "Standard Test Method for Sulphur in Petroleum and Petroleum Products by Energy Dispersive X-Ray Fluorescence Spectrometry", provides rapid and precise measurement of total sulphur in petroleum and petroleum products with a minimum of sample preparation. A typical analysis time is 1–5 minutes per sample.

The quality of many petroleum products is related to the amount of sulphur present. Knowledge of sulphur concentration is necessary for processing purposes. There are also regulations promulgated in national and international standards that restrict the amount of sulphur present in some fuels.

The test covers the determination of total sulphur in petroleum and petroleum products that are single phase and either liquid at ambient conditions, liquefiable with moderate heat or soluble in hydrocarbon solvents. These materials can include diesel fuel, jet fuel, kerosene, other distillate oil, naphtha, residual oil, lubricating base oil, hydraulic oil, crude oil, unleaded gasoline, gasoline–ethanol blends, biodiesel and similar petroleum products.

The concentrations of substances in the test method were determined by the calculation of the sum of the mass absorption coefficients times the mass fraction of each element present. This calculation was made for dilutions of representative samples containing approximately 3% of interfering substances and 0.5% sulphur. For samples with high oxygen contents (>3% wt), sample dilution as described in the test method or matrix matching must be performed to ensure accurate results.

Interlaboratory studies on precision revealed the scope to be 17 mg/kg to 4.6 mass percent. An estimate of this test method's pooled limit of quantitation (PLOQ) is 16.0 mg/kg as calculated by the procedures in practice ASTM D6259. However, because instrumentation covered by this test method can vary in sensitivity, the applicability of the test method at sulphur concentrations below approximately 20 mg/kg must be determined on an individual basis. An estimate of the limit of detection is three times the reproducibility standard deviation, and an estimate of the limit of quantitation is 10 times the reproducibility standard deviation.

Samples containing more than 4.6 mass percent sulphur can be diluted to bring the sulphur concentration of the diluted material within the scope of the test method.

Samples that are diluted can have higher errors than indicated in Section 16 of the test method than non-diluted samples.

A fundamental assumption in the test method is that the standard and sample matrices are well matched, or that the matrix differences are accounted for. Matrix mismatch can be caused by C/H ratio differences between samples and standards or by the presence of other heteroatoms.

9.2.16 Metals Contents

The metals contents of new lubricants (and the additives they contain) and lubricants in use are best measured individually by methods such as atomic absorption (AA), inductively coupled plasma (ICP) spectroscopy, or x-ray fluorescence (XRF). The metals contents of base oils can also be tested, to determine if they have been contaminated during transportation or storage.

The old techniques of wet chemical precipitation, filtering and weighing have been superseded except for reference purposes by modern analytical equipment which uses comparative instrumental techniques. These relate the test sample readings to those given by standard calibration material. Some methods can use oil as received, in others it needs to be diluted with a solvent, while some require more complex sample preparation techniques.

At one time, flame photometry (ASTM D3340) was a general-purpose technique used for many elemental analyses in chemical laboratories, but now its use is largely confined to older laboratories and for the rapid measurement of sodium and lithium contents found in greases and some lubricating oils. The sample can be diluted with solvent, or for greater sensitivity can be ashed and dissolved in water. In either case, the liquid is aspirated into a flame. Characteristic wavelengths of light are emitted in the flame depending on the elements present, the light is dispersed by a spectrometer and the relevant wavelengths are measured electronically. The intensity of emission from the sample is compared with standards of known elemental concentration and the concentrations in the sample thereby estimated.

AA spectroscopy (ASTM D4628) has largely replaced flame photometry for routine analysis in many petroleum laboratories. A monochromatic beam of light, characteristic of the element to be measured, is produced by a hollow-cathode lamp containing the element. This light is passed through an intense flame, such as from an acetylene/nitrous oxide burner, and the sample in solvent solution is aspirated into the flame. Atoms of the element being measured are produced in the flame and absorb the characteristic radiation, and the drop in its intensity can be measured and related to the elemental content of the sample using calibration standards. Elements routinely measured by AA include barium, calcium, magnesium and zinc.

Flame photometry is a form of emission spectroscopy, but the term is usually applied to more complex apparatus employing an electric arc or a plasma to excite the spectral lines from the sample, rather than a flame. The higher-energy excitation source gives higher sensitivity and produces more useful spectral lines for the purpose of analysis, which can help to overcome problems of interferences between elements.

ICP emission spectroscopy (ASTM D4951) is a more recent technique of emission spectroscopy, tending to be less sensitive than arc systems for some elements but normally giving greater accuracy. A stream of ionisable gas such as argon is excited by a powerful radio frequency coil and is converted into a plasma. The sample in a solvent is sprayed into the plasma and characteristic spectral lines are emitted. The method has moderate sensitivity for barium and phosphorus but more for calcium, magnesium, sulphur and zinc. It can also be used for trace metals in used oil analysis, when ASTM D5185 is used.

XRF spectroscopy (ASTM D4927) has become a very widely used analytical technique of high speed and good precision. Sample preparation is not normally needed, the oil being placed in a small cell with a thin plastic film window at the bottom. This is attached to a vacuum chamber, and the sample is irradiated through the window with powerful x-rays. The sample emits secondary x-rays whose wavelengths are characteristic of the elements present. These are analysed with a crystal diffraction spectrometer, and the intensity at selected wavelengths is measured with an x-ray counter (for example, a Geiger counter). The method is sensitive for most elements but is unsuitable for those lighter than silicon and therefore excludes sodium and magnesium. The precision of the method is relatively good and can sometimes be further improved by extending the counting time (the time taken to measure the emitted x-rays). There are considerable interelement interference effects, and correct calibration is essential for accurate analysis.

ASTM D5185, "Standard Test Method for Multi-Element Determination of Used and Unused Lubricating Oils and Base Oils by Inductively Coupled Plasma Atomic Emission Spectrometry (ICP-AES)", covers the rapid determination of 22 elements in used and unused lubricating oils and base oils. It provides rapid screening of used oils for indications of wear. Test times approximate a few minutes per test specimen, and detectability for most elements is in the low milligram per kilogram range. In addition, the test covers a wide variety of metals in virgin and re-refined base oils.

9.2.17 Hydrocarbon Type Analysis

The ASTM D3238 method covers the calculation of the carbon distribution and ring content of olefin-free petroleum oils from measurements of refractive index, density and molecular weight (n-d-M). The composition of complex petroleum fractions is often expressed in terms of the proportions of aromatic rings (R_A), naphthene rings (R_N) and paraffin chains (Cp) that would comprise a hypothetical mean molecule. Alternatively, the composition may be expressed in terms of a carbon distribution, that is, the percentage of the total number of carbon atoms that are present in aromatic ring structures (% Ca), naphthene ring structures (% Cn) and paraffin chains (% Cp).

The refractive index and density of the oil are determined at 20°C. The molecular weight is determined experimentally or estimated from measurements of viscosity at 100°F and 210°F (37.8°C and 100°C). These data are then used to calculate the carbon distribution (% Ca, % Cn, % Cp) or the ring analysis (R_A, R_N) using the appropriate set of equations.

The carbon distribution and ring content serve to express the gross composition of the heavier fractions of petroleum. These data can be used as an adjunct to the bulk properties in monitoring the manufacture of lubricating base oils by distillation and solvent refining or hydroprocessing, and in comparing the composition of different base oils from different crude oils. Additionally, the data can often be correlated with critical product performance properties.

9.3 TESTS FOR ADDITIVES

A wide variety of additives can be distinguished visually by colour and apparent visual viscosity when viewed in a glass container. The key tests on additives are:

- Appearance.
- Viscosity.
- Elemental contents/sulphated ash.
- H_2S.
- Infrared spectrum.

As well as the possible supply of an incorrect grade, appearance will also indicate whether the additive has been contaminated by water, in which case it may appear milky and form a noticeable sediment on standing. Viscosity will vary somewhat from batch to batch of the same material but should lie within a reasonable band for correctly manufactured material.

Infrared absorption is a more sophisticated test than visual appearance and can give much information about the consistency of an additive package composition. However, an agreement between the supplier and customer on how results will be interpreted is most desirable.

9.3.1 METALS AND NON-METALS CONTENTS

See Sections 9.2.15 and 9.2.16.

9.3.2 SULPHATED ASH

The sulphated ash of a lubricating oil is the residue, in percent by weight, remaining after three processes: burning the oil, treating the initial residue with sulphuric acid and burning the treated residue. It is a measure of the non-combustible constituents (usually metallic materials) contained in the oil.

New, straight mineral lubricating oils contain essentially no ash-forming materials. Many of the additives used in lubricating oils, such as detergents, contain metallo-organic components, which will form a residue in the sulphated ash test, and so the concentration of such materials in an oil is roughly indicated by the test. Thus, during manufacture, the test gives a simple method of checking to ensure that the additives have been incorporated in approximately the correct amounts. However, since the test combines all metallic elements into a single residue, additional testing may be necessary to determine if the various metallic elements are in the oil in the correct proportions.

Test methods IP 163 or ASTM D874 (which are technically equivalent) are used to determine the sulphated ash content of an additive or additive-containing lubricant. The ash content of a base oil (which should be free of ash-forming additives) can be determined using test methods IP 4 or ASTM D482.

9.3.3 Infrared Spectroscopy

With the availability of simple inexpensive infrared absorption spectrophotometers, it is becoming increasingly common for purchasers of base oils or additives to demand a reference sample of the material to be supplied, and to take an infrared spectrum of this sample. It is unlikely to provide much detailed information on the composition of the sample, it but can be used as a "fingerprint" to identify the material.

With moderate interpretative skills, the spectrum can be used to identify which of several known additives or oils is under test, and if the composition of these has been changed. However, both base oils and additives can show changes in their absorption spectra, depending on processing conditions or raw material sources, without there being necessarily a change in quality or performance. Use of fingerprinting can persuade suppliers to discuss such changes with customers rather than supply without declaration of a change. Some purchasers will reject deliveries of oil or additives if the infrared spectrum of the new product does not match that of a retained reference sample.

Spectral changes are always worthy of discussion with a supplier, with whom there needs to be an understanding of possible and permissible variations. On the other hand, significant quality changes could take place and not show up in an infrared spectrum. An example of the infrared spectrum of a lubricant is shown in Figure 9.2.

9.4 SPECIFICATIONS FOR RAW MATERIALS

The quality control process starts with the agreement of specifications for the purchased raw materials: base oils and additives. When these were agreed, lines of communication for rapid discussion of quality problems concerning incoming material should have been set up, and samples of representative on-grade material obtained for reference purposes. More than a few millilitres of each product is required, for these samples will be used to visually check incoming material and also as references when there is some doubt about quality. Suppliers should also be asked for their own certificates of analysis to be supplied with the material when it is delivered, so that their inspections can be compared with the agreed specifications and with test results as they are obtained in the blending plant laboratory.

As well as financial and stock drawing arrangements, the purchasing agreements should include specifications for all the components to be used in the blends, with provision of samples of each material for reference purposes. This applies to all base oil and additive suppliers. It may be that viscosity modifier, pour depressant and performance package additives are supplied by different additive companies, and if

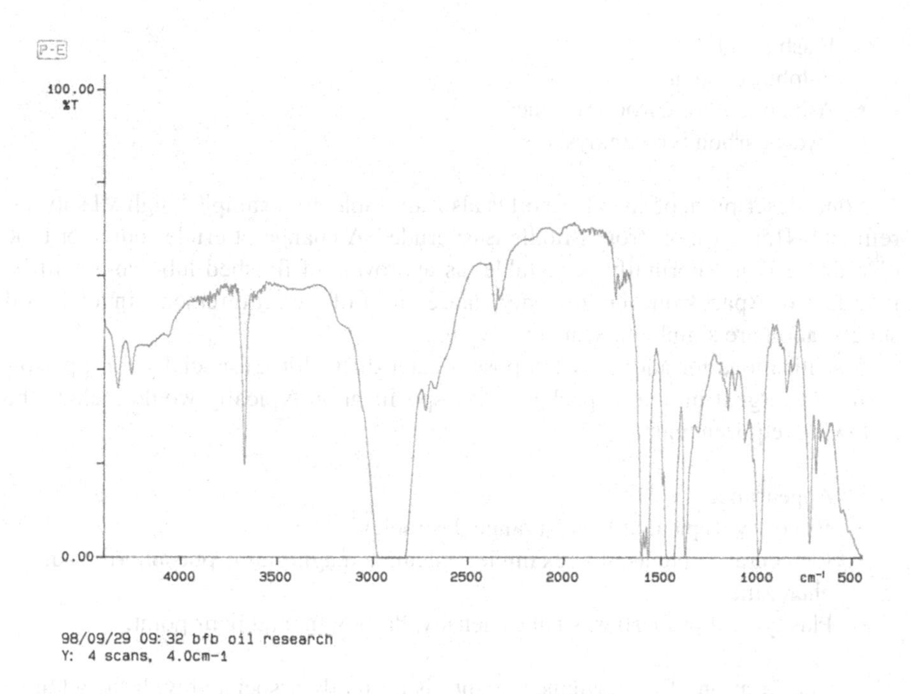

FIGURE 9.2 Example infrared spectrum. (Source: Pathmaster Marketing Ltd.)

the oil has a part synthetic content, then this will probably be supplied by a different company from the main base oil provider.

The reference samples are very important and should be stored in a cool, dark place. They are used for monitoring future deliveries of the materials, and as reference materials in any dispute with the supplier. If a laboratory infrared spectrophotometer is available, then infrared traces should be obtained from each sample. Infrared spectra are commonly used as "fingerprints" by which commercial deliveries of both additives and finished lubricants can be compared with reference samples. (In large blending plants with sophisticated laboratories, the use of Fourier transform infrared spectrophotometers with built-in computers permits automated comparison of spectra.)

Proper purchase and supply specifications must be drawn up with all component suppliers. For base oils, the following is a representative list of tests that might be found in a typical purchase specification:

- Appearance (clear and bright).
- Colour.
- Density.
- Viscosity.
- Viscosity index.
- Pour point.

- Flash point.
- Sulphur content.
- Ash content or carbon residue.
- Hydrocarbon type analysis.

Some description of each base oil is also advisable, for example "high VI solvent-refined SN150 base oil from Middle East crude". A change of crude source or type of refining is not normally acceptable, as approvals of finished lubricants will be invalidated. Apart from the viscosity clause, most of the requirements in a base oil specification are simple maxima or minima.

Specifications for additives can pose greater difficulties, especially for a performance (detergent inhibitor) package. The specification typically would include the following requirements:

- Appearance.
- Viscosity (typical at least, a range desirable).
- Elemental contents, for example, calcium, magnesium, phosphorus, sulphur, zinc.
- Plus typical properties, such as density, flash point and pour point.

The declaration of performance quality is normally associated with the additive code or name and is contained within the purchase contract rather than the purchase specification.

For viscosity improvers and pour depressants, minimum performance targets for a dilute blend in a specified test oil should be specified. There should be a general description of the purpose and quality level provided by the additive package, and guidelines on the correct handling of the materials should always be sought. Many additive packages require heating to reduce viscosity before they can be pumped into a blending vessel, and overheating may easily decompose the additive, possibly with the production of toxic fumes. It is the additive supplier's responsibility to provide handling and storage guidance and information on the hazards and toxicity of his products in some form of safety data package.

The agreement of specification limits for the elemental contents between the additive supplier and the lubricant blender may take a little time. The supplier has limitations on the control of manufacturing that can be achieved, and testing will add additional variability to the measured quality of the material supplied. The lubricant blender has testing variability, but also needs additional flexibility to allow for the inevitable quality variations of his own manufacturing. The finished oil must, however, be manufactured to a specification that ensures that the finished oil is of the quality claimed, and hence the purchase specification between the lubricant blender and additive supplier is squeezed by the output specifications of the two parties.

A satisfactory agreement must be made if business is to proceed smoothly. If the two parties have not previously worked together, it may be necessary to have a period of cooperative sample exchange and analysis in order to arrive at common ground on the analysis of the elements of the package. Experience has indicated that this can be done, but considerable time may elapse before such discussions are successfully concluded and a satisfactory specification is developed.

9.5 SUMMARY

Many tests are available to assess the properties and performance of both base oils and additives used as raw materials with which to blend finished lubricants. Because a blending plant is another link in the entire supply chain, the ability of the blending plant staff to manufacture finished lubricants of the correct quality depends in part on the ability of the suppliers of base oils and additives to deliver the appropriate-quality raw materials.

10 Testing and Analysis of Blended Lubricants

10.1 INTRODUCTION

In the author's opinion, the only true test of a lubricant's performance is whether it functions satisfactorily for several years in the machinery for which it was formulated. Lubricant specifications, laboratory tests, rig tests and field trials are only useful as a guide or prediction of likely performance in service. In assessing the possible suitability of a specific lubricant for a specific application, it is critically important to remember to check that the predictive tests are the best ones for that application.

The range of tests is huge, but tests may be conveniently separated into four categories:

- Laboratory tests.
- Bench and rig tests.
- Engine tests.
- Field tests.

All these tests attempt to shorten the time taken to assess the properties or performance of lubricants, because the "real-life" performance of lubricants can take many years to establish. If actual performance in service was the only method by which lubricants could be evaluated, the development of new or improved products could take a very long time indeed. It is, however, very important to remember that laboratory, bench, rig, engine and even field tests are useful mainly for establishing lubricants that are unlikely to perform satisfactorily in practical operations, not those that definitely will perform satisfactorily for many years.

Laboratory tests, usually for physical or chemical properties, are usually quick and comparatively inexpensive. Bench and rig tests, which are intended to assess performance properties, such as anti-wear, corrosion inhibition or deposit formation, are performed in specially designed equipment or machines that are smaller and/or less complex than real engines or gearboxes. Consequently, they tend to be shorter and cheaper than engine tests. Engine tests are performed in special buildings, to try to provide as much repeatability and reproducibility as possible, and are more realistic, but are longer and more expensive than laboratory or rig tests. Field tests try to reproduce real-life operating conditions but are lengthy and expensive.

However, for a blending plant, many of the tests used to assess the characteristics and performance of finished lubricants are not appropriate or suitable for the purposes of quality control or quality assurance. Thus, this chapter focuses only on those tests that are required in a blending plant.

10.2 LABORATORY TESTS FOR LUBRICANTS

A multitude of physical and chemical tests yield useful information on the characteristics of lubricating oils. However, the quality or performance features of a lubricant cannot be adequately described on the basis of physical and chemical tests alone. Thus, major purchasers of lubricating oils, such as the military, equipment builders and many commercial consumers, include performance tests as well as physical and chemical tests in their purchase specifications. Physical and chemical tests are of considerable value in maintaining uniformity of products during manufacture. (They may also be applied to used oils to determine changes that have occurred in service, and to indicate possible causes for those changes.)

Some of the most commonly used tests for physical or chemical properties of lubricating oils are outlined in the following sections, with brief explanations of the significance of the tests from the standpoint of the refiner and consumer. Detailed information on methods of test are contained in handbooks issued by the organisations listed in Table 9.1.

- The American Society for Testing and Materials (ASTM) Annual Standards for Petroleum Products and Lubricants.
- The Energy Institute (formerly the Institute of Petroleum [IP]) Standard Methods for Testing Petroleum and Its Products.
- U.S. Federal Test Method Standard No. 791.
- International Organization for Standardization (ISO).
- Comité Européen de Normalisation (CEN).
- Deutsche Institut für Normung (DIN).
- American National Standards Institute (ANSI).
- Association Français de Normalisation (AFNOR).
- Japanese Automotive Standards Organisation (JASO).

A wide number of tests are performed to assess the physical or chemical properties of lubricants. Tests for physical properties include

- Kinematic viscosity (capillary viscometer, low shear).
- Low-temperature viscosity:
  Brookfield viscometer (pumpability).
  Cold cranking simulator (CCS) (crankability).
  Mini-rotary viscometer (MRV) (pumpability).
- High-temperature viscosity:
  Tapered bearing simulator (high temperature, high shear).
  Ravenfield viscometer.
- Pour point.
- Flash point:
  Pensky–Martens closed cup (PMC) (for contamination).
  Cleveland open cup (COC) (for composition).
- Volatility:
  NOACK.
  Gas chromatography (simulated distillation).

Air jet.
Distillation.
- Foaming tendency and stability.
- Density (specific gravity).

Tests for chemical properties include:

- Acidity or alkalinity:
  Acid number (AN).
  Neutralisation number.
  Base number (BN).
- Sulphated ash.
- Elemental analysis:
  Flame photometry.
  Atomic absorption (AA) spectroscopy.
  Inductively coupled plasma (ICP) emission spectroscopy.
  X-ray fluorescence (XRF) spectroscopy.
- Infrared (IR) spectroscopy.

Bench and rig tests are more commonly used to evaluate the potential performance of industrial lubricants. However, because of the increasing complexity and expense of engine tests, they are being used increasingly to evaluate automotive lubricants. Bench and rig tests used for lubricants include:

- Oxidation resistance:
  IP 280.
  IP 306.
  ASTM D943.
  ASTM D2893.
  PDSC (ASTM D6186).
  TEOST (ASTM D6335).
- Thermal stability:
  Panel coker (FTM 3462).
- Shear stability:
  Diesel injector (IP 294).
  Tapered roller bearing.
- Corrosion resistance:
  Steel (IP 135, ASTM D665).
  Copper (IP 154, ASTM D130).
  High-temperature corrosion bench test (ASTM D5968).
- Anti-wear, load carrying or extreme-pressure properties:
  Timken (ASTM D2782).
  Falex (ASTM D2670).
  Four-ball ASTM D2783).
  FZG (IP 351).
- Cam and tappet rigs.

Again, details of the methods and equipment used for each of these tests may be found in the reference books and publications issued by the testing authorities listed in Table 9.1. Many of the laboratory tests have been described in Chapter 9 on testing base oils and additives.

There are some bench and rig tests for finished lubricants that may be of relevance to blending plant quality control and quality assurance.

10.2.1 Oxidation Resistance

Oxidation of lubricating oils depends on the temperature, the amount of oxygen contacting the oil and the catalytic effects of metals. If the oil's service conditions are known, these three variables can be adjusted to provide a test that closely represents actual service. However, oxidation in service is often an extremely slow process, so the test may be time-consuming. To shorten the test time, the test temperature is usually raised and catalysts added to accelerate the oxidation. Unfortunately, these measures tend to make the test a less reliable indication of expected field performance. As a result, very few oxidation tests have received wide acceptance, although a considerable number are used by specific laboratories that have developed satisfactory correlations for them.

Many tests have been developed over the years to measure aspects of the oxidation performance of different types of oils. These include:

- ASTM D943: TOST (turbine oil stability test).
- IP280: CIGRE test.
- DIN 51596: PNEUROP test.
- ASTM D2272: RPVOT (rotating pressure vessel oxidation test).
- ASTM D4742: TFOUT (thin-film oxygen uptake test).
- CEC L-48–95: DKA test.
- ASTM D4310: Modified ASTM D943 test.
- ASTM D2893: Oxidation and thermal stability test.
- ASTM D6186: PDSC (pressure differential scanning calorimetry) test.

However, many of these tests are much too long to be of any practical value to a lubricant blending plant. For example, the TOST (ASTM D943) can run for up to 10,000 hours. Even the modified TOST (ASTM D4310) takes 1000 hours to run, and although the PNEUROP test takes only 24 hours, it is not really applicable to base oils, having been developed to assess high-performance air compressor oils.

Two oxidation stability tests that are suitable for use in blending plants are the RPVOT (ASTM D2272, previously called the RBOT [rotating bomb oxidation test]) and the PDSC test (ASTM D6186).

ASTM D2272, "Standard Test Method for Oxidation Stability of Steam Turbine Oils by Rotating Pressure Vessel", which is technically equivalent to the IP 229 test, was widely known as the RBOT. The test uses an oxygen-pressured cylinder to evaluate the oxidation stability of new and in-service oils in the presence of water and a copper catalyst coil at 150°C. This procedure is often used as a screening test and quality control test because of the speed at which results can be obtained. A 50 g sample of the test oil, 5 ml of water and a copper wire catalyst are placed in a small cylinder and pressurised to 90 psi with oxygen at room temperature (25°C). The cylinder is then placed in a 150°C bath and rotated at 100 rpm. The pressure increases as the vessel is heated, reaches a maximum value and then drops as oxidation occurs. Once a 25 psi drop from the maximum pressure is observed, the amount of time from the vessel being placed in the bath through the drop is reported. An unmodified oil typically will run less than 30 minutes, and a high-quality formulated oil can run in excess of 1000 minutes (16.67 hours).

The test is not intended to be a substitute for ASTM D943 or to be used to compare the service lives of new oils of different compositions. It can also be used to assess the remaining oxidation test life of an in-service oil.

Appendix X1 of ASTM D2272 describes a new optional turbine oil (unused) sample nitrogen purge pretreatment procedure for determining the percent residual ratio of the RPVOT value for the pretreated sample divided by the RPVOT value of the new (untreated) oil, sometimes referred to as a "percent RPVOT retention". This nitrogen purge pretreatment approach was designed to detect volatile antioxidant inhibitors that are not desirable for use in high-temperature gas turbines.

The ASTM D6186 PDSC test method is used to determine the oxidation induction time of lubricating oils subjected to oxygen at 3.5 MPa (500 psig) and temperatures between 130°C and 210°C. The test method is faster than other oil oxidation tests and requires a very small amount of sample. A 3 mg sample of test oil is weighed in a new sample pan and placed in a test cell. The cell is heated at a rate of 100°C/min to a specified test temperature and then pressurised with oxygen to 3.5 MPa. The pressure is maintained using a flow rate of 100 ml/min. The test is run for 120 minutes or until after the oxidation exotherm has occurred. The oxidation induction time is defined as the time from when the oxidation valve is opened to the onset time for the oxidation exotherm. The onset time is extrapolated from the thermal curve. If more than one oxidation exotherm is observed, then the largest exotherm is reported.

Another PDSC test is CEC L-85–99, in which a sample of candidate oil is heated in a pressure differential scanning calorimetry unit to a defined temperature and then held isothermally at that temperature for up to 2 hours to determine the oxidative induction time (OIT). In this test, 2 mg of sample is heated between 50°C and 210°C and then held at that temperature for up to 2 hours in a closed system at 100 psi (6.9 bar) overpressure. The OIT, expressed in minutes, is the onset time observed from achieving the isothermal temperature. The test was developed for use in the ACEA E5 heavy-duty diesel engine oil specification, in which the straight pass/fail criterion is set at 65 minutes for candidate oils.

10.2.2 Antioxidant Content

ASTM D6971, "Standard Test Method for Measurement of Hindered Phenolic and Aromatic Amine Antioxidant Content in Non-Zinc Turbine Oils by Linear Sweep

Voltammetry", covers the determination of hindered phenol and aromatic amine antioxidants in new or used non-zinc turbine oils. The test measures concentrations from 0.0075% wt up to those found in new oils by measuring the amount of current flow at a specified voltage in the produced voltammogram. The test is not designed or intended to detect all the antioxidant intermediates formed during the thermal and oxidative stressing of the oils, which are recognised as having some contribution to the remaining useful life of the used or in-service oil. Nor does it measure the overall stability of an oil, which is determined by the total contribution of all species present. Before making a final judgement on the remaining useful life of the used oil, which might result in the replacement of the oil reservoir, it is advised to perform additional analytical techniques in accordance with ASTM D6224, ASTM D4378 and ASTM D2272, having the capability of measuring the remaining oxidative life of the used oil.

The test is applicable to non-zinc-type turbine oils as defined by ISO 6743 Part 4, Table 1. These are refined mineral oils containing rust and oxidation inhibitors but not anti-wear additives. The test is also suitable for manufacturing control and specification acceptance.

A measured quantity of sample is dispensed into a vial containing a measured quantity of acetone-based electrolyte solution and a layer of sand. When the vial is shaken, the hindered phenol and aromatic amine antioxidants and other solution-soluble oil components in the sample are extracted into the solution, and the remaining droplets suspended in the solution are agglomerated by the sand. The sand–droplet suspension is allowed to settle out and the hindered phenol and aromatic amine antioxidants dissolved in the solution are quantified by voltammetric analysis. The results are calculated and reported as mass percent of antioxidant or as millimoles of antioxidant per litre of sample for prepared and fresh oils, and as a percent of the remaining antioxidant for used oils.

When a voltammetric analysis is obtained for a turbine oil inhibited with a typical synergistic mixture of hindered phenol and aromatic amine antioxidants, there is an increase in the current of the produced voltammogram between 8 and 12 s (or 0.8 and 1.2 V applied voltage) for the aromatic amines, and an increase in the current of the produced voltammogram between 13 and 16 s (or 1.3 and 1.6 V applied voltage) for the hindered phenols in the neutral acetone test solution. Hindered phenol antioxidants detected by voltammetric analysis include, but are not limited to, 2,6-di-tert-butyl-4-methylphenol, 2,6-di-tert-butylphenol and 4,4'-methylenebis (2,6-di-tert-butylphenol). Aromatic amine antioxidants detected by voltammetric analysis include, but are not limited to, phenyl alpha naphthylamines and alkylated diphenylamines.

10.2.3 Thermal Stability

Thermal stability, as opposed to oxidation stability, is the ability of an oil or additive to resist decomposition under prolonged exposure to high temperatures with minimal oxygen present. Decomposition may result in thickening or thinning of the oil, increasing acidity, the formation of sludge or any combination of these.

Thermal stability tests usually involve static heating, or circulation over hot metal surfaces. Exposure to air is usually minimised, but catalyst coupons of various metals may be immersed in the oil sample. While no tests have received wide acceptance, a number of proprietary tests are used to evaluate the thermal stability of products such as hydraulic fluids, gas engine oils and oils for large diesel engines.

10.2.4 Rust Protection

The rust-protective properties of lubricating oils are difficult to evaluate. Rusting of ferrous metals is a chemical reaction that is initiated almost immediately when a specimen is exposed to air and moisture. Once initiated, the reaction is difficult to stop. Thus, when specimens are prepared for rust tests, extreme care must be taken to minimise exposure to air and moisture so that rusting will not start before the rust-protective agent has been applied and the test begun. Even with proper precautions, rust tests do not generally show good repeatability or reproducibility.

Most laboratory rust tests involve polishing or sandblasting a test specimen, coating it with the oil to be tested and then subjecting it to rusting conditions. Testing may be in a humidity cabinet, by atmospheric exposure or by some form of dynamic test. In the latter category is ASTM D665, "Rust Preventive Characteristics of Steam Turbine Oil in the Presence of Water". Method IP 135 is technically equivalent to ASTM D665. In this test, a steel specimen is immersed in a mixture of distilled or synthetic seawater and the oil under test. The oil and water moisture is stirred continuously during the test, which usually lasts for 24 hours. The specimen is then examined for rusting. A photograph of ASTM D665/IP135 test specimens is shown in Figure 10.1.

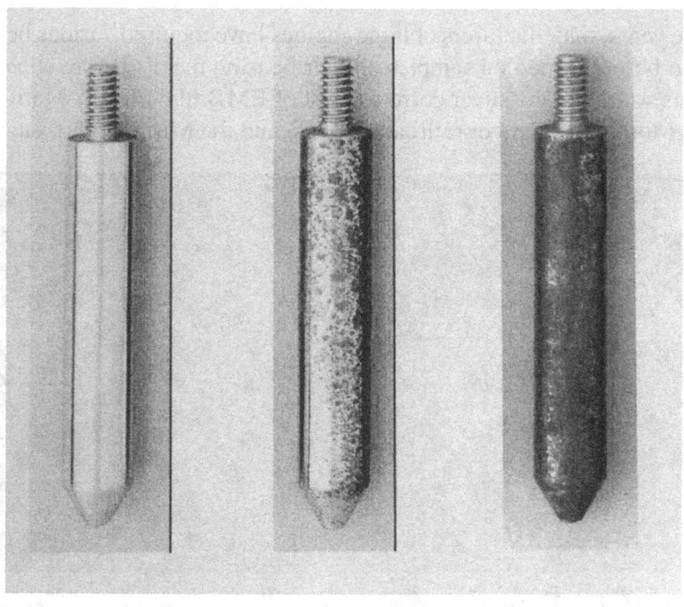

FIGURE 10.1 ASTM D665/IP 135 steel corrosion test specimens.

The IP 220 "Emcor" method is used widely in Europe for oils and greases. This procedure employs actual ball bearings, which are visually rated for rust on the outer races at the end of the test. This is a dynamic test that can be run with a given amount of water flowing into the test bearing housings. Severity may be increased with the use of salt water or acid water in place of the standard distilled water.

10.2.5 Corrosion Resistance

The most commonly used test is the copper strip corrosion test (ASTM D130, IP 154), in which a polished copper strip is immersed in the sample and heated, and its colour and condition after the test are compared with a defined classification scheme. For lubricating oils, the test is usually performed at a temperature of 100°C for 3 hours. A photograph of ASTM D130/IP 154 test specimens is shown in Figure 10.2. The classification of the copper strip after completion of the test is

- 0 Initial polished copper strip
- 1 Slight tarnish: 1a light orange, almost the same as the freshly polished strip; 1b dark orange
- 2 Moderate tarnish: 2a claret red; 2b lavender; 2c multicoloured, with lavender blue or silver or both overlaid on claret red; 2d silvery; 2e brassy or gold
- 3 Dark tarnish: 3a magenta overcast on brassy strip; 3b multicoloured with red and green showing, but no grey
- 4 Corrosion: 4a transparent black, dark grey or brown with peacock green barely showing; 4b graphite or lustreless black; 4c glossy or jet black

Over the years, manufacturers of large engines have required various bench corrosion tests to be performed on samples of their bearing metals before approving oils. Typical tests would be the silver corrosion test of EMD (the Electro-Motive Division of General Motors, builders of railroad engines) and the Mirrlees corrosion test.

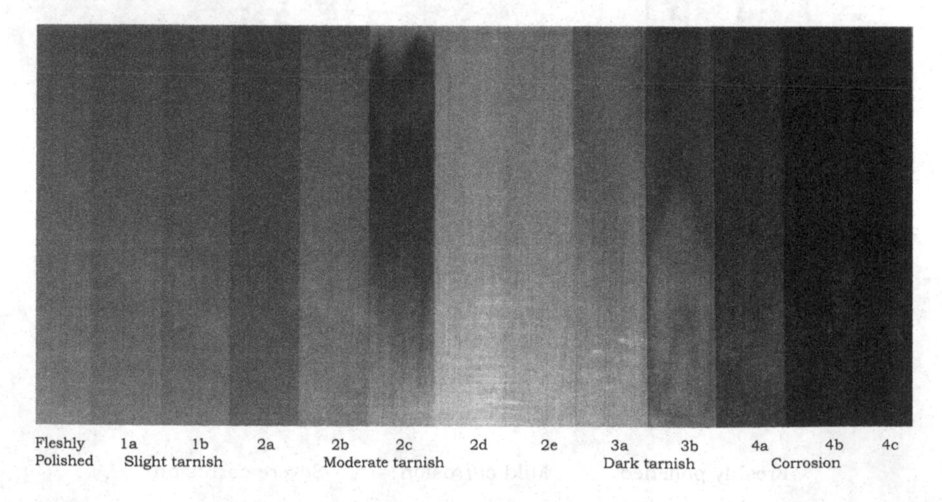

FIGURE 10.2 ASTM D130/IP 154 copper corrosion test specimens.

10.2.6 Shear Stability

Non-Newtonian fluids exhibit a lower viscosity when subjected to high shear rates than Newtonian fluids having the same viscosity at low shear. Oils which contain viscosity index improving polymers exhibit non-Newtonian viscosity behaviour. This reduced viscosity at high shear rate may be temporary and reversible, depending on the stability to shear of the VI improving polymer. If, however, the shear rate is too high, the polymer molecules may be broken.

When the polymer molecules are subjected to high shear rates, the available energy causes the polymer coils to elongate and squeeze through narrow orifices without the polymer itself being changed. As the stress is removed, the polymer recovers its original configuration and the level of viscosity at low shear remains unchanged. This type of reversible viscosity loss is often called "temporary viscosity loss" and is seen in tests such as the tapered bearing simulator.

In high-temperature and/or high-energy situations, and particularly where localised oil shear is much higher than that of the average within the volume of oil, the polymer will undergo a permanent change by rupture of the molecular chains. Oil flow which induces extension of the polymer molecules, and cavitation phenomena, can also cause polymer breakdown.

When a long molecule breaks into two shorter molecules, the overall thickening effect is reduced and a permanent loss in the oil viscosity results. The longer the polymer chain, the higher the oil temperature, the greater the shearing action and the greater the tendency for rupture to occur. The greatest shearing effect occurs in areas of thin-film lubrication at high sliding speeds, particularly in oil pumps, gearboxes, valve gear and piston ring zones.

The level of permanent viscosity loss depends on the molecular weight and structure of the polymer used to thicken the oil. Different samples of polymers can therefore be assigned a shear stability index (SSI) from which the performance, in terms of in-service viscosity loss, can be estimated.

Various types of apparatus have been used to measure SSI, including motored gearboxes, sonically agitated systems and various systems of contra-rotating discs/paddles. The most common technique, however, is to use a diesel injector pump to spray the oil through a diesel injector nozzle, recycling the oil so that it undergoes many passes through the nozzle. In the nozzle, the oil is subjected to high-speed thin-film changes of direction and is then atomised. Various diesel injector rigs and test methods are used, but in Europe the Kurt Orbahn rig (test method IP 294), named after the apparatus manufacturer, uses a Bosch injector pump and is specified for most multigrade oil specification testing. The test measures the percent viscosity loss at 100°C, after 30 passes through the diesel injector nozzle, of fluids that contain a viscosity index improving polymer. A photograph of a diesel injector test rig is shown in Figure 10.3. The Coordinating European Council (CEC) has adopted the IP 294 test method, as CEC L-14–93. The DIN 51382 test method is technically equivalent to IP 294.

The ASTM has also adopted IP 294, as test method ASTM D6278, "Standard Test Method for Shear Stability of Polymer-Containing Fluids Using a European Diesel Injector Apparatus". The ASTM notes that the test is used for quality control

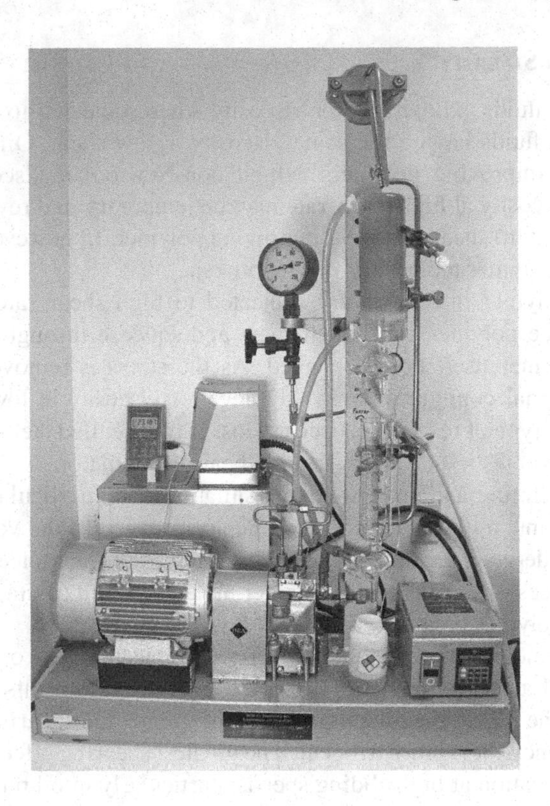

FIGURE 10.3 Diesel injector test rig.

purposes by manufacturers of polymeric lubricant additives and their customers. It is not intended to predict viscosity loss in field service in different equipment under widely varying operating conditions, which may cause lubricant viscosity to change due to thermal and oxidative changes as well as by the mechanical shearing of polymers. However, when the field service conditions cause mainly mechanical shearing of the polymer, there may be a correlation between the results from this test method and results from the field. ASTM also notes that while the test method uses test apparatus as defined in CEC L-14–93, it differs from it in the period of time required for calibration.

ASTM D7109 is a similar test that measures shear stability but uses both 30 and 90 passes through the diesel injector nozzle. Test method ASTM D5275, "Standard Test Method for Fuel Injector Shear Stability Test (FISST) for Polymer-Containing Fluids", is a similar test using a slightly different fuel injector nozzle. Like the other two tests, it is intended to measure shear stability in the absence of thermal or oxidative effects on polymer degradation. ASTM D5275 was originally published as Procedure B of ASTM D3945. The FISST method was made a separate test method after tests of a series of polymer-containing fluids showed that Procedures A and B of ASTM D3945 often give different results.

Another test that can be used to determine the shear stability of viscosity index improving polymers uses the FZG gear rig, method IP 351. In this test, special gear wheels are run in the lubricant under test in a dip lubrication system at a constant speed for a fixed time. The bulk oil temperature is controlled and the loading of the gear teeth is set according to the chosen condition. At the end of the test period, the oil is assessed for permanent viscosity loss and the difference in viscosity between the new and sheared oil is used to characterise the shear stability of the oil.

The FZG rig has for some years been used to evaluate the shear stability of automotive engine oils. The test has been found to be particularly relevant to applications in which the engine and transmission share a common lubricant. The test includes operating conditions appropriate to engine oils, gear oils and hydraulic fluids. The test is likely to be particularly suitable for oils that exhibit high shear stability, when the diesel injector shear stability test (IP 294) may not be sufficiently severe. A photograph of the FZG gear rig is shown in Figure 10.4.

Another method for measuring shear stability is ASTM D5621, "Standard Test Method for Sonic Shear Stability of Hydraulic Fluids". The test was developed using ASTM D2603. It permits the evaluation of shear stability with minimum interference from thermal and oxidative factors that may be present in some applications. It has been found to be applicable to fluids containing both readily sheared and shear-resistant polymers, and correlation with performance in the case of hydraulic applications has been established, particularly with regard to polymer shear experienced in a hydraulic vane pump test procedure. Correlation with performance for polymer-containing automotive engine oils has not been established.

ASTM D5621 evaluates the shear stability of a hydraulic oil containing a viscosity index improving polymer in terms of the permanent loss in viscosity that results from irradiating a sample of the oil in a sonic oscillator. The test measures the kinematic viscosity at 40°C before and after irradiation and reports the resulting percentage loss in viscosity. The test can be useful in predicting the continuity of this property in an oil where no change is made in the base oil or the polymer. It is not intended to predict the performance of polymer-containing oils in service. It has also

FIGURE 10.4 FZG test rig.

been found that there is little or no correlation between the shear degradation results obtained by means of sonic oscillation and those obtained by mechanical devices. Also, the sonic test may rate different families of polymers in a different order than that obtained from mechanical devices. The ASTM has made no detailed attempt to correlate the results by the sonic and diesel injector methods.

The IP 294 test method is likely to be more suitable for use in a lubricant blending plant than the IP 351 test method, as the former is usually easier and quicker to run. However, larger blending plants may need to have both tests, as the latter is useful for assessing the load-carrying properties of gear oils and transmission fluids.

10.2.7 Anti-Wear and Extreme-Pressure Tests

One of the main functions of a lubricant is to reduce mechanical wear. Closely related to wear reduction is the ability of lubricants of the extreme-pressure type to prevent scuffing, scoring and seizure as applied loads are increased. As a result, a considerable number of machines and procedures have been developed to try to evaluate anti-wear and extreme-pressure properties. In a number of cases, the same machines are used for both purposes, although different operating conditions are usually used.

Wear can be divided into four classifications based on the cause: abrasive wear, corrosive (chemical) wear, adhesive wear and fatigue wear.

Abrasive wear is caused by abrasive particles, either contaminants carried in from outside or wear particles formed as a result of adhesive wear. In either case, oil properties do not have much direct influence on the amount of abrasive wear that occurs, except through their ability to carry particles to filtering systems that remove them from the circulating oils.

Corrosive or chemical wear results from chemical action on the metal surfaces combined with rubbing action that removes corroded metal. A typical example is the wear that may occur on cylinder walls and piston rings of diesel engines burning high-sulphur fuels. The strong acids formed by combustion of the sulphur can attack the metal surfaces, forming compounds that can be fairly readily removed as the rings rub against the cylinder walls. Direct measurement of this wear requires many hours of test unit operation, which is often done as the final stage of testing new formulations. However, useful indications have been obtained in relatively short periods of time by means of sophisticated electronic techniques.

Adhesive wear in lubricated systems occurs when, owing to load, speed or temperature conditions, the lubricating film becomes so thin that opposing surface asperities can make contact. If adequate extreme-pressure additives are not present, scuffing and scoring can result, and eventually seizure may occur. Adhesive wear can also occur with extreme-pressure lubricants when the reaction kinetics of the additives with the surfaces are such that metal-to-metal contact is not fully controlled.

A number of machines, such as the Almen, Alpha LFW-1, Falex, four-ball, SAE, Timken and Optimol SRV, are used to determine the loading conditions under which seizure, welding or drastic surface damage to test specimens would be permitted. Of these, the Alpha LFW-1, four-ball, Falex, Timken and Optimol SRV are also used to measure wear at loads below the failure load. The results obtained with these machines do not necessarily correlate with field performance, but in some cases, the results from certain machines have been reported to provide useful information for specific applications. Photographs of a four-ball machine and a friction and wear test machine are shown in Figures 10.5 and 10.6, respectively.

Either actual machines or scale model machines are used in testing the anti-wear properties of lubricants. Anti-wear hydraulic fluids are tested for anti-wear properties in pump test rigs using commercial hydraulic pumps of the piston-and-vane type. Frequently, industrial gear lubricants are tested in the FZG gear machine (IP 334), a scale model machine that reasonably approximates commercial practice with respect to gear tooth loading and sliding speeds. Correlation of the FZG test with field performance has been good.

Fatigue wear occurs under certain conditions when the lubricating oil film is intact and metallic contact of opposing asperities is either nil or relatively small. Cyclic stressing of the surfaces causes fatigue cracks to form in the metal, leading to fatigue spalling or pitting. Fatigue wear occurs in rolling element bearings and gears when there is a high degree of rolling, and adhesive wear, associated with sliding, is negligible. A variation of the fatigue pitting wear is the micropitting wear mechanism that results in small pits in the surfaces of some gears and bearings. Asperity interaction and high loads as well as metallurgy are factors influencing micropitting. An upgraded version of the standard FZG spur gear tester is used to evaluate gear oil micropitting performance according to a German Research Institute test method called FVA (Forschungsvereinigung Antriebstechnik) Method 54.

The understanding of the influence of lubricant composition and characteristics is an evolving technology. Some products are currently available that address fatigue-induced wear in anti-friction bearings and gears caused by the presence of small amounts of water in the Hertzian load zones. In these instances, the fatigue is called water-induced fatigue, and oil chemistry can reduce the negative effects of water in

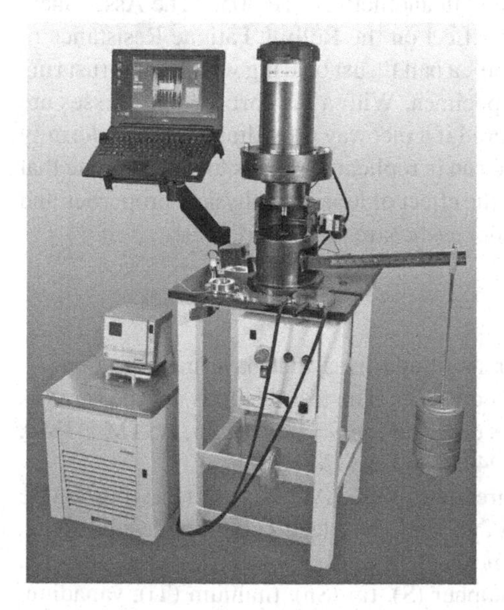

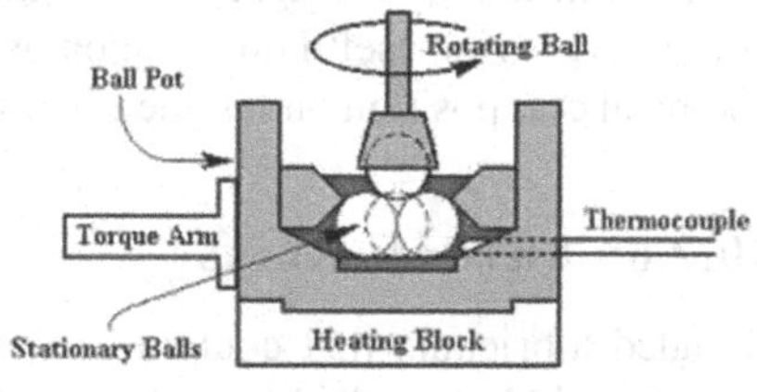

FIGURE 10.5 Four-ball test machine and schematic. (From Savant Labs. With permission.)

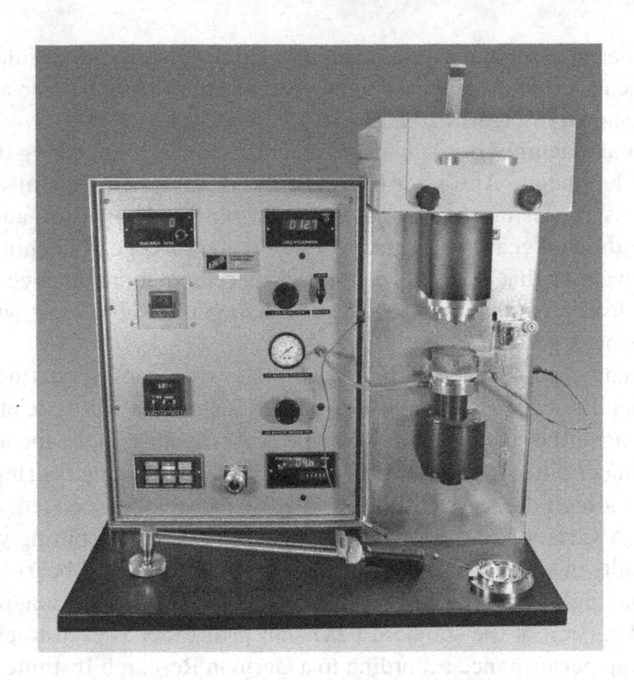

FIGURE 10.6 Friction and wear test machine. (From Savant Labs. With permission.)

the load zones of anti-friction bearings and gears. Two tests have been introduced by the Institute of Petroleum for studies of this type. In IP 300, "Pitting Failure Tests for Oils in a Modified Four-Ball Machine", oils are tested under conditions that cause cyclic stressing of steel bearing balls. To shorten the test time, the load and stress are set higher than usual for the ball bearings. In another test, IP 305, "The Assessment of Lubricants by Measurement of Their Effect on the Rolling Fatigue Resistance of Bearing Steel Using the Unisteel Machine", a ball thrust bearing with a flat thrust ring is used. The flat thrust ring is the test specimen. With a flat thrust ring, stresses are higher than would be normally encountered if a raceway providing better conformity were machined in the ring. Both tests are run in replicate. Results to date indicate that both tests provide useful information on the effect of lubricant physical properties and chemical composition on fatigue life under cyclic stressing conditions.

10.2.8 Metals Contents

Blended lubricants that contain one or more additives that have metallic components should be checked for their metals contents. The test methods are the same as those used for base oils and additives, as described in Section 9.2.16, ASTM D3340, ASTM D4628, ASTM D4951, ASTM D4927 and ASTM D5185.

The elements covered by the tests are aluminium (Al), barium (Ba), boron (B), calcium (Ca), chromium (Cr), copper (Cu), iron (Fe), lead (Pb), magnesium (Mg), manganese (Mn), molybdenum (Mo), nickel (Ni), phosphorous (P), potassium (K), silicon (Si), silver (Ag), sodium (Na), sulphur (S), tin (Sn), titanium (Ti), vanadium (V) and zinc (Zn).

Blending plant engineers and supervisors will find it difficult to appreciate the comparatively low levels of precision found in chemical analysis of petroleum products compared with physical measurements within their own disciplines. There are many reasons for this, but the most important are:

- Precise basic standards can be defined for most measurements of length or distance, for example, a bar of metal or the wavelength of light. Concentration standards are, however, much more difficult to set up.
- Where concentration standards can be produced, for example, by adding weighed amounts of a pure substance to a diluent, such samples often have limited application.
- This is because in most cases the standard samples need to be quite similar to the unknown samples for analysis. In spectrographic analytical methods, the different elements interfere mutually with each other's output signals, and the base matrix can also have a considerable influence.
- For classical wet chemical analysis, precipitated material can be lost, or false precipitates of other interfering elements can be unknowingly produced.
- Spectroscopic devices have considerable "noise" in both the excitation and detection areas, which is significant in relation to the magnitude of the signals being measured and limits the sensitivity at low concentration levels.
- Petroleum samples are normally liquids which are not completely stable either chemically or physically.

Repeatability of spectrographic methods for lubricating oil analysis lies in the range of 2% to 5% of the mean value, while estimates of reproducibility from cooperative test programmes show values ranging from around 4% to 20%. This does not mean that two laboratories working together, such as a lubricant supplier and a lubricant customer, cannot come to considerably closer agreement for a specific analysis. The problem of establishing meaningful specifications for new lubricants depends on close cooperation between the supplier and the user. Accuracy depends on the quality of the reference or calibration standards used.

10.3 TESTS FOR BLENDING PLANT QUALITY CONTROL OF SPECIFIC TYPES OF LUBRICANTS

As noted previously, not all the tests described above are required for quality control in lubricant blending plants. However, the tests used in a blending plant are different for different types of lubricant, because some are more important than others. In each of the sections below, tests described as critical are definitely required to control final product quality. Tests described as important can be done but are not needed to achieve good-quality control, while tests described as desirable can be done if the laboratory has the necessary equipment and there is sufficient time to deliver the results before the final product is packaged or shipped.

The most important criterion for any of these tests is that they can be completed quickly and the results communicated (using the blending plant's distributed control system [DCS]) to supervisors and operators on the shop floor. This is discussed in more depth in Section 10.4.

10.3.1 Automotive and Industrial Engine Oils

For gasoline and diesel engine oils, whether automotive or industrial and on highway or off highway, the critical tests for quality control are:

Kinematic viscosity at 40°C and 100°C.
Viscosity index.
Dynamic viscosity at –5°C–40°C, depending on viscosity grade.
High-temperature, high-shear viscosity.
Sulphated ash content.
Zn, Ca, Mg, P and S contents, depending on the specific formulation.

These tests should be considered as mandatory. Other important tests that could be run, time permitting, are:

Base number.
NOACK volatility or evaporative loss.
Foaming tendency.
One desirable test for multigrade gasoline and diesel engine oils is shear stability.

10.3.2 Automotive and Industrial Gear Oils

For automotive and industrial gear oils, whether mineral oil, polyalphaolefin, polyisobutene or polyalkylene glycol based, the critical tests in a blending plant laboratory are:

Kinematic viscosity at 40°C and 100°C.
Viscosity index.
Pour point.
Foaming tendency.
Demulsibility.
S and P contents.

Important tests for gear oils are rust protection (IP135/ASTM D665) and corrosion resistance (IP 154/ASTM D130). If time permits, two desirable tests are four-ball anti-wear and FZG load carrying.

10.3.3 Automatic Transmission Fluids

The range of tests for new automatic transmission fluids is vast, but for a lubricant blending plant the critical tests needed to assess product quality are:

Kinematic viscosity at 40°C and 100°C.
Viscosity index.

Dynamic viscosity at −25°.
High-temperature, high-shear viscosity.
Foaming tendency.
Air release value.

In a lubricant blending plant, important tests for automatic transmission fluids are:

Rust protection.
Corrosion resistance.
Four-ball anti-wear.

If the blending plant has the necessary test equipment, a desirable test for automatic transmission fluids is the friction clutch test, although this test has not been described in this book.

10.3.4 Hydraulic Oils

The critical tests for hydraulic oils in a lubricant blending plant depend on whether the product is monograde or multigrade. For both monograde and multigrade hydraulic oils, the critical tests are:

Kinematic viscosity at 40°C and 100°C.
Viscosity index.
Pour point.
Flash point.
Foaming tendency.
Demulsibility.
Air release value.
Zn, P and/or S contents.

Important tests for both monograde and multigrade hydraulic oils are:

Acid number.
Four-ball anti-wear.
Rust protection.
Corrosion resistance.

A critical test for multigrade hydraulic oils only is shear stability, using either the diesel injector rig test (IP 294) or the FZG shear stability test (IP 351).

10.3.5 Turbine Oils

For steam turbine and industrial gas turbine oils, critical tests in a lubricant blending plant are:

Kinematic viscosity at 40°C and 100°C.
Viscosity index.

Pour point.
Flash point.
Foaming tendency.
Demulsibility.
Rust protection.
Corrosion resistance.
P and S contents.

For both oils, important tests are acid number and air release value. Again, if time permits a useful desirable test is the four-ball anti-wear test.

10.3.6 Compressor Oils

For air compressor oils, reactive gas compressor oils and natural gas compressor oils, the critical tests for quality control in a lubricant blending plant are:

Kinematic viscosity at 40°C and 100°C.
Viscosity index.
Pour point.
Flash point.
Emulsification tendency.

Important tests for all three types of compressor oils include acid number, rust protection and corrosion resistance. A desirable test is foaming tendency.

The tests required for compressor oils will depend on whether the product is an air compressor oil, a reactive gas compressor oil, a natural gas compressor oil or a refrigerator compressor oil. The former may benefit from having its DIN 51506 oxidation stability tested, while the latter will probably require its thermal stability (particularly in the presence of a refrigerant) to be assessed.

10.3.7 Metalworking Fluids

The tests required for metalworking fluids will depend on whether they are neat oil types or water-mix types and the type of metalworking or metal forming for which they are intended. Critical tests are:

Kinematic viscosity at 40°C.
Rust protection.
Corrosion resistance.
Emulsion stability (for water-mix fluids).

Four-ball anti-wear and extreme-pressure characteristics are important for neat metalworking fluids, while foaming tendency is an important test for water-mix metalworking fluids. A desirable test for both types is elemental contents.

10.4 PROCESSING AND COMMUNICATION OF TEST RESULTS

Quality control tests in a blending plant must be comparatively quick, usually taking no more than an hour or so. They should also be comparatively cheap in terms of test materials and reagents. (The equipment used may be quite expensive.)

Management of blending should include a system for planning when quality control samples will be taken and delivered to the laboratory; the laboratory has to be able to plan its workload too.

Quality control tests for each blend must be done simultaneously, not sequentially, so the elapsed time between sampling and delivery of the results is as short as possible. Test results should be available to the blending unit operators or supervisors as soon as possible after a blend has been completed. Many of the laboratory tests used for quality control in blending plants have now been automated. Examples include kinematic viscosity, pour point, flash point and density.

With modern automated blending and quality management systems in blending plants, test results can be input directly by laboratory staff to the DCS. The results are therefore available instantly to blending unit operators and supervisors.

However, a mechanism needs to be set up to alert the operators and supervisors that the results for their current blend are available, so they do not have to stare at a computer screen waiting for the results to arrive.

10.5 SUMMARY

Numerous tests are used to assess the characteristics and performance of lubricants, but most of these tests apply to the use of these products in mechanical equipment.

Tests used for quality control purposes in a blending plant need to be relatively quick and to not involve complex mechanical equipment. The results of tests should be available either while a blend is being completed or within a few hours of the blend being transferred to a holding tank, but certainly before the product is required to be packaged or shipped.

The tests used in a blending plant will vary between product types; some tests are more critical than others for certain products.

Mechanisms need to be set up to ensure that the blending units and the laboratory communicate with each other quickly and effectively.

11 Lubricant Product Quality Control

11.1 INTRODUCTION

The *Oxford English Dictionary* defines *quality* as "the standard of something as measured against other things of similar kind" or "general excellence". The dictionary also defines *quality control* as "a system of maintaining standards in manufactured products by testing a sample of the output against the specification".

Among many definitions of *quality*, *Webster's International Dictionary* defines *quality* as "peculiar or essential character", "a distinctive inherent feature" and "degree of conformance to a standard (as of a product or workmanship)".

All these definitions are historic, generic and conventional. More importantly, they are of little practical value when trying to either control or guarantee the quality of a product or service in a business environment.

11.2 DEFINITION OF *QUALITY*

In business, engineering and manufacturing, quality has a pragmatic interpretation as the non-inferiority or superiority of something. Quality is a perceptual, conditional and somewhat subjective attribute and may be understood differently by different people. Consumers may focus on the specification quality of a product or service, or how it compares to competitors in the marketplace. Producers might measure the conformance quality, or degree to which the product or service was produced correctly.

Numerous definitions and methodologies have been created to assist in managing the quality-affecting aspects of business operations. Many different techniques and concepts have evolved to improve product or service quality. There are two common quality-related functions within a business:

- *Quality assurance:* The prevention of defects, such as by the deployment of a quality management system (QMS) and preventative activities like failure mode effect analysis (FMEA).
- *Quality control*: The detection of defects, most commonly associated with testing which takes place within a QMS, typically referred to as verification and validation.

The ISO 9000 standard (explained in more detail in Chapter 14) defines quality as the "degree to which a set of inherent characteristics fulfils requirements".

The American Society for Quality describes quality as a subjective term for which each person has his or her own definition. In technical usage, quality can have two meanings:

- "The characteristics of a product or service that bear on its ability to satisfy stated or implied needs".
- "A product or service free of deficiencies".

The first of these definitions is that used by the ISO 8402:1986 standard.

The common element of the business definitions is that the quality of a product or service refers to the perception of the degree to which the product or service meets the customer's expectations. Quality has no specific meaning unless related to a specific function and/or object.

Another word, *reliability*, also needs to be defined correctly. In answer to the familiar question "Why do you buy a Ford car?" the words *quality* and *reliability* often come back. The two words are used synonymously, often in a totally confused way. Clearly, part of the acceptability of a product or service will depend on its ability to function satisfactorily over a period of time. It is this aspect of performance which is given the name *reliability*.

Reliability is the ability of the product or service to continue to meet the customer's requirements. Reliability ranks with quality in importance, since it is a key factor in many purchasing decisions where alternatives are being considered. Many of the general management issues related to achieving product or service quality are also applicable to reliability.

It is important to realise that the "meeting the customer requirements" definition of *quality* is not restrictive to the functional characteristics of products or services. Anyone with children knows that the quality of some of the products they purchase is more associated with satisfaction in ownership than some functional property. This is also true of many items, from antiques to certain items of clothing. The requirement for status symbols accounts for the sale of some executive cars, bank accounts, charge cards and even hospital beds! The requirements are of paramount importance in the assessment of the quality of any product or service, and quality is the most important aspect of competitiveness.

The ability to meet the customer requirements is vital, not only between two separate organisations, but within the same organisation. There exists in every department, every office, even every household, a series of suppliers and customers. A personal assistant is a supplier to his or her manager; is he or she meeting his or her requirements? Does he or she receive error-free typing set out as and when he or she wants it? If so, then the organisation has a quality typing service. Does an airline steward (male or female) receive from the airline's catering company the correct food trays in the right quantity?

Throughout and beyond all organisations, whether they be manufacturing companies, banks, retail stores, universities or hotels, there is a series of quality chains which may be broken at any point by one person or one piece of equipment not meeting the requirements of the customer, internal or external. In many cases, this failure usually finds its way to the interface between the organisation and its outside customers. The people who operate at that interface, such as the air steward, usually experience the ramifications.

A great deal is written and spoken about employee motivation as a separate issue. In fact, the key to motivation and quality is for everyone in the organisation to have well-defined customers (an expansion of the word) beyond the outsider that actually purchases or uses the ultimate product or service. It needs to apply to anyone to whom an individual gives a part, service or information, in other words, the results of his or her work.

Quality has to be managed; it will not just happen. Clearly, it must involve everyone in the process and be applied throughout the organisation. Some people in customer organisations never see, experience or touch the products or services that their companies purchase, but they do see things like invoices. If every fourth invoice from a certain supplier carries at least one error, what image of quality is transmitted?

Failure to meet the requirements in any part of a quality chain has a way of multiplying, and failure in one part of the system creates problems elsewhere, leading to yet more failure, more problems and so on. The price of quality is the continual examination of the requirements and the ability to meet them. This alone will lead to a "continuing improvement" philosophy. The benefits of making sure that the requirements are met at every stage, every time, are truly enormous in terms of increased competitiveness and market share, reduced costs, improved productivity and delivery performance, and the elimination of waste.

11.3 MEETING CUSTOMER REQUIREMENTS AND QUALITY CONTROL

If quality is meeting the customer's requirements, this has wide implications. The requirements may include availability, delivery, reliability, maintainability and cost-effectiveness, among many other benefits or features. On the list of things to do, the first one is to find out what are the requirements.

If an organisation is dealing with a supplier–customer relationship crossing two organisations, then the supplier must establish a "marketing" activity charged with this task. The marketing staff must, of course, understand not only the needs of the customer but also the ability of their own organisation to meet the demands. If a customer places a requirement on a person to run a mile (ca. 1500 m) in 4 minutes, in almost all cases the person (supplier) knows that he or she will be unable to meet this demand, unless something is done to improve his or her running performance. Of course, the person may never be able to achieve this requirement.

Within organisations, between internal customers and suppliers, the transfer of information regarding requirements frequently varies from poor to completely absent. How many executives rarely bother to find out what are the requirements of their internal customers (their personal assistants)? Can their handwriting be read? Do they leave clear instructions? Do the personal assistants always know where the manager is? Equally, do the personal assistants establish what their managers need, such as accurate filing, timely meeting schedules, clear messages or a tidy office?

These internal supplier–customer relationships are often the most difficult to manage in terms of establishing the requirements.

To achieve quality throughout an organisation, each person in the quality (supply) chain must question every interface.

11.3.1 Customers

- Who are my immediate customers?
- What are their true requirements?
- How do or can I find out what the requirements are?
- How can I measure my ability to meet the requirements?
- Do I have the necessary capability to meet the requirements? (If not, then what must change to improve the capability?)
- Do I continually meet the requirements? (If not, then what prevents this from happening, when the capability exists?)
- How do I monitor changes in the requirements?

11.3.2 Suppliers

- Who are my immediate suppliers?
- What are my true requirements?
- How do I communicate my requirements?
- Do my suppliers have the capability to measure and meet the requirements?
- How do I inform them of changes in the requirements?

The answers to all these questions form the basis for starting quality control, quality assurance and quality management methodologies and actions.

11.4 PROCEDURES FOR CONTROL OF QUALITY

Quality management is the subject of Chapter 14, so this chapter focuses on methodologies for controlling quality, particularly as applied to the blending of lubricants.

A quality manual documents an organisation's QMS. It can be a paper manual or an electronic manual. According to ISO 9001 Section 4.2.2, an organisation's quality manual should:

- Define the scope of the QMS.
- Explain reductions in the scope of the QMS.
- Justify all exclusions (reductions in scope).
- Describe how the QMS processes interact.
- Document the quality procedures or refer to them.

Note that an ISO 9001 quality manual does *not* simply reproduce the ISO 9001 standard. While this is a common practice, it not only fails to comply with the standard but also fails to serve any useful function. (It should also be noted that ISO 9004:2015 does not require a quality manual.)

Quality planning involves setting quality objectives and then specifying the operational processes and resources that will be needed to achieve those objectives. Quality planning is one part of quality management. A quality plan is a document that is used to specify the procedures and resources that will be needed to carry out a project, perform a process, realise a product or manage a contract. Quality plans also specify who will do what and when.

An organisation's quality policy defines senior management's commitment to quality. A quality policy statement should describe an organisation's general quality orientation and clarify its basic intentions. Quality policies should be used to generate quality objectives and should serve as a general framework for action. Quality policies can be based on the ISO 9000 quality management principles and should be consistent with the organisation's other policies.

A quality objective is a quality-oriented goal. A quality objective is something an organisation aims for or tries to achieve. Quality objectives are generally based on or derived from the organisation's quality policy and must be consistent with it. They are usually formulated at all relevant levels within the organisation and for all relevant functions.

11.4.1 Checking Raw Materials

For a blending plant, the amount of inspection depends on the source of the raw material and how it is delivered. For example, base oils delivered by pipeline from an adjacent base oil manufacturing plant and owned by the same organisation may require very little checking. However, base oils supplied by sea will require careful inspection to ensure that they are the correct materials and that they have arrived uncontaminated. Additives pose particular problems because they are relatively difficult to test and analyse, yet the supplier may be remote and not in a close relationship with the blending plant. Supply may be via an agent, and the product could have spent some time in storage, possibly not under ideal conditions.

The quality control process starts with the agreement of specifications for the purchased products as discussed in Chapter 9. When these were agreed, lines of communication for rapid discussion of quality problems concerning incoming material should have been set up, and samples of representative on-grade material obtained for reference purposes. More than a few millilitres of each product are required, as these samples will be used to check incoming material visually and also as references when there is some doubt about quality. Suppliers should also be asked for their own certificates of analysis to be supplied with the material when it is delivered, so that their inspections can be compared with the agreed specifications and to test results as they are obtained in the blending plant laboratory.

Tests on base oil samples should follow those given in specifications, but not every test may need to be done. The following may be considered:

- Appearance.
- Viscosity.
- Flash point.
- Pour point.

- Index.
- Density.
- Ash content.

The above tests are ranked in order of urgency and in terms of the information that they may give. Appearance may immediately indicate if the oil being received is very different from the sample that was provided by the supplier, which should be viewed alongside the new material. If the oil is noticeably hazy, then it is possible that it is contaminated, possibly by water. If water is suspected, a "crackle" test can be performed, but this is not an entirely reliable indication of water content. The viscosity will indicate if the correct grade has been supplied, and the flash point will tell if the material has been contaminated by light products during shipping or pipelining operations. The pour point may vary somewhat between shipments but should always be in the same area; it can be useful for distinguishing between paraffinic and naphthenic base oils. The viscosity index (VI) will likewise confirm whether a specified quality of high-VI base oil has been supplied, and the density can indicate if the base stock has been made from a different crude than was originally the case. Finally, ash content is another measure of possible contamination, possibly by additive materials.

In most cases, contamination by additives or fuel products would be detected by the appearance test. It is suggested that the first three tests be performed on material during delivery, and the full spectrum of tests is performed on a full tank of oil after delivery has taken place. Of course, if any adverse indications are seen, pumping should stop immediately until the quality has been investigated further and the cause of the discrepancy ascertained. Initial delivery into a holding tank is desirable, with product being transferred to the main bulk tank after testing.

Additives may be delivered in bulk or in drums, and if in bulk, preliminary tests should be performed on the delivering tanker (sea, rail or road) before any material is offloaded. With the cargo should travel samples from the additive plant as the material was loaded, and these and material from the delivery should be compared with the stock reference samples. A wide variety of additives can be distinguished visually by colour and apparent visual viscosity when viewed in a glass container. The key tests on additives are:

- Appearance.
- Viscosity.
- Elemental contents.
- Sulphated ash.
- Hydrogen sulphide (H_2S) smell.

Infrared absorption is a more sophisticated test than visual appearance and can give much information about the consistency of additive package composition. However, an agreement between the supplier and customer on how results will be interpreted is most desirable.

As well as the possible supply of an incorrect grade, appearance will indicate whether the additive has been contaminated by water, in which case it may appear milky and form a noticeable sediment on standing. Viscosity will vary somewhat from batch to batch of the same material but should lie within a reasonable band for correctly manufactured material. The key parameters of an additive are usually the elemental contents, such as calcium, magnesium, phosphorus, zinc and sulphur. These are best measured individually by such methods as atomic absorption, inductively coupled plasma spectroscopy or x-ray fluorescence, but if such equipment is not available, then a carefully performed sulphated ash content will normally provide sufficient reassurance.

Additives such as viscosity improvers and pour point depressants will have to be checked by dissolving a small proportion in a base oil and measuring the viscosity and pour point of the blends. In relation to supply specification and discussion of potency of such products with the supplier, it will be necessary to use a standardised test oil for making the blends and not stock base oil from tankage.

Additives will in most cases be supplied in drums, and it is essential that at all times these are stored correctly. Additive manufacturers will give guidance on drum storage and also on the storage of additives in bulk. Drums should not be stored on their ends because of the possibility of water collecting in the drum head around the bung, and detergent additives in particular must be stored so that they do not become overheated. Performance package additives stored in dark drums in tropical sunshine are very likely to decompose or at least lose their efficiency.

Drums, like other packages, are normally sampled by the statistical rule which indicates that the number of packages examined must be equal to the cube root of the total. In this connection, it should be noted that every batch of material contained within one delivery should be treated as different material.

When drums are opened for sampling, note any pungent odours from the barrels, particularly if these are unusual. *Do not sniff at the bunghole after removing the bung.* If the additive has partially decomposed due to overheating, it may be evolving H_2S, which is extremely toxic, as well as malodorous compounds, and should be treated as hazardous. On all deliveries of detergent inhibitors or performance packages it is worthwhile to run a simple test for H_2S liberation in the laboratory and compare this with results obtained on the standard samples. To do this, a small amount of additive is placed in a flask and put in a boiling water bath, and a filter paper moistened with lead acetate solution is placed on top. If the additive is decomposing, the paper will rapidly blacken. Some additives may give slight colouration after 5 minutes, and this can be ignored.

11.4.2 Controlling Quality during Blending

If on-grade raw materials are used, then production control should consist simply of ensuring that the correct amounts of the raw materials have been correctly mixed together. The base oils have to be in the correct proportions, the correct percentages of viscosity improver and pour depressant additives have to be incorporated, and the right amount of detergent or other additives have to be included so that the oil meets its performance targets.

Each batch or consolidation of batches should be fully tested in the case of manually operated plants, while for automated blending where accuracy has been fully demonstrated, statistical sampling may be employed. For any type of blending, particular vigilance must be observed after a change of oil grade or type.

The tests to be used during blending will depend on the type of oil being blended. Test for engine oils will be different than those for hydraulic fluids, gear oils or turbine oils (See Chapter 10).

For automotive engine oils, the following blending control tests are recommended:

Monograde Oil	Multigrade Oil
Viscosity (100°C)	Viscosity (100°C)
Pour point	Pour point
Elements or ash	Elements or ash
	CCS or MRV

The viscosity is obviously a very important property, but it must be remembered that for multigrade oils, a low or high viscosity can be due to either a misblend of the base oils or an incorrect amount of viscosity improver. The pour point can normally be adjusted by extra additions of pour point depressant if it is initially deficient. To determine if the correct amount of detergent additive has been added, a carefully performed sulphated ash is acceptable, although analytical determination of at least one of the elements (calcium, magnesium and so on) is desirable. For multigrade oils, it is desirable that some form of low-temperature test other than the pour point is performed. Low-temperature specification testing is very time-consuming, but consideration should be given to setting up a cold cranking simulator (CCS) or mini-rotary viscometer (MRV) apparatus. (Strictly, the CCS is a requirement for certifying SAE W–grade conformity.)

Another possibility is to have a non-specification test, such as kinematic viscosity at –20°C, compared with a reference blend which has been made in the laboratory and thoroughly checked against the final oil specification. This latter reference blend is desirable anyway to use as a standard for the tests, and possibly for providing to a bulk purchaser of the oil. For additive content checks, side-by-side determinations of the sulphated ash on a reference oil, as well as the oil being blended, would be one way of increasing the value of this test.

For industrial oils, elements other than those identifying detergents will need to be checked. These include zinc, phosphorus, sulphur and possibly chlorine or other elements. Other quick tests, such as foaming performance, air release, kinematic viscosity at 40°C and 100°C (to check VI) and flash point, may need to be done. Purchasers of finished lubricants may require the infrared spectrum to be consistent batch to batch. Checks using an infrared spectrophotometer may therefore be needed to ensure that this is so, before oil is released for sale.

Due to the difficulties of test duration, repeatability and standardisation, it is not recommended that laboratory performance tests (such as the four-ball extreme-pressure test, oxidation stability or corrosion resistance) be used for production control purposes.

Unlike many other manufactured products in other industries, quality control in lubricant blending is relatively easy to achieve because many of the laboratory tests (particularly kinematic viscosity, pour point, flash point and elemental analysis) are quick and have been automated. Blends can be checked quickly immediately after blending (or toward the end of mixing), and the results can be available before the blend is pumped to a holding tank or storage tank.

It is not proposed here to go into detail on control chart procedures for the production of lubricants, but this is always a worthwhile procedure. They should include features such as the results of first-time blends and the numbers of reblends required to bring material into grade. Consistent errors such as undercharging of viscous additives can be identified and corrected, and a more consistent product will be produced more efficiently. Comprehensive literature exists on statistical quality control techniques.

11.4.3 Testing Finished Products

Some purchasers of lubricants may require the oil supplier to provide a certificate of quality that the specific product meets the specification requirements, whether for an established national or international specification or a company's own specification.

In addition to testing each blend to determine whether the process has been completed successfully, when a blend is added to a holding tank or a storage tank that already contains the same product, that tank's contents will need to be mixed and the mixture tested. In general, unless a blend is passed directly from a holding tank to filling and packaging or to bulk delivery by barge, rail tank wagon or road tanker, the storage tank containing the final finished product will also need to be tested for quality control.

11.4.4 External Monitoring Systems

Oil companies and their blending plants should be aware that, as well as customers who purchase lubricants in bulk, other organisations may test lubricants which have been provided by the blending plant.

The most prominent of these is the U.S. Engine Oil Licensing and Certification Scheme (EOLCS), which was set up to administer the award of the American Petroleum Institute (API) licence. Oils bearing the API symbol are taken from the field in a random manner and analysed for conformity with the original qualified formulation. Both physical properties and elemental content are examined, and guidelines are laid down as to the permitted variation before the supplier is called in to account for any differences. Manufacturers or marketers of oils that fail to meet their claimed specifications are asked to recall and correct the products. In a few cases of repeated failures or non-conformity, manufacturers' or marketers' API licences are revoked, meaning that the oil(s) in question may no longer be sold. Lubricant marketing companies that apply for an API Certification Mark are required to pay an annual fee, which varies depending on the volume of engine oil sold, to fund the EOLCS system.

In Europe, discussions have been taking place for some time to decide whether a similar scheme should be introduced for product verification in the marketplace. The United Kingdom Lubricants Association (UKLA) introduced a new initiative, the Verification of Lubricants Specifications (VLS), in September 2013, aimed at protecting consumers. VLS is an independent organisation providing a credible and trusted means to verify lubricant specifications.

Member companies of VLS who suspect that a competitor's product may not meet its claimed performance specification are able to raise a complaint. VLS then responds to complainants by independently and confidentially investigating and checking compliance to industry standards and performance claims within the UK marketplace. It also engages with market participants and relevant authorities and industry bodies to pursue incorrect or false product quality and standards, with focus being placed on education and/or undertaking of remedial action to improve compliance. The company whose product is being investigated has a confidential "right of reply", but where remedial or corrective action is not undertaken, VLS will provide relevant information to industry bodies and relevant authorities to take necessary action. VLS aims to raise market awareness through publicly reporting findings from its investigations on the quality of lubricants sold in the market and promoting the importance of lubricant quality and performance specifications within every UK lubricant market.

At the time of writing, neither the Association Technique de l'Industrie Européene des Lubrifiants (ATIEL) nor the European Union of Independent Lubricant Manufacturers (UEIL) had announced plans to introduce a system similar to EOLCS in Europe. In Europe, lubricant marketing companies are required, under the ATIEL Code of Practice, to self-certify the quality of their products. Aspects of this, using the ISO 9000 standard, will be discussed further in Chapter 14.

It has been suggested in some countries that trading standards officers may examine oils on the market and request justification of their quality claims from the oil blender and/or the additives supplier. Performance tests would not be envisaged, but they could check for physical properties and for the correct amount of the declared additive being present in the oils on the market. Such action has already taken place in cases where members of the public have complained about apparent inadequacies in the quality of lubricants which they have purchased.

11.4.5 Component and Formulation Codes

A very effective method of maintaining control of lubricant formulations, from confidentiality, quality assurance and quality control viewpoints, is to use component and formulation codes.

Each raw material from each supplier is given a unique code. All product formulations are also given unique codes. For example, antioxidant 1 from supplier A is assigned the code AO-01 and antioxidant 2 from supplier A is assigned code AO-02.

If antioxidant 1 can also be obtained from supplier B, it should be given a different code, such as AO-05. Each base oil should also be given a unique code.

Code letters for each type of base oil and additive should be different. An example scheme is

100 SN Group I base oil	BO-01
150 SN Group I base oil	BO-02
300 SN Group I base oil	BO-03
500 SN Group I base oil	BO-04
150 BS Group I base oil	BO-05
100 N Group II base oil	BO-11
150 N Group II base oil	BO-12
220 N Group II base oil	BO-13
500 N Group II base oil	BO-14
4 cSt Group III base oil	BO-21
6 cSt Group III base oil	BO-22
8 cSt Group III base oil	BO-23

Of course, in the above scheme, all the Group I base oils, the Group II base oils and the Group III base oils are from the same supplier(s). If a 150 SN Group I base oil is purchased from a different supplier, it may need to be given the code BO-06, unless its properties are almost identical to those of BO-02. If the properties are not almost identical, product formulations may need to be adjusted slightly to compensate for the differences.

Similar styles of code can be used for synthetic base oils, polyalphaolefins (PAOs), diesters (DIEs), polyol esters (POEs), polyisobutenes (PIBs) and polyalkylene glycols (PAGs). Additives can be coded as follows:

Viscosity index improvers	VI
Detergents	DE
Dispersants	DS
Pour point depressants	PP
Oxidation inhibitors	AO
Anti-wear additives	AW
Extreme-pressure additives	EP
Corrosion inhibitors	CI
Metal passivators	MP
Demulsifiers	DM
Emulsifiers	EM
Friction modifiers	FM
Anti-foam additives	AF
Biocides	BI

When it comes to additive packages, care should be taken to make these different from those for single additives:

Dispersant inhibitor pack	DIpack
Gear oil pack	GOpack
Hydraulic fluid pack	HFpack
Compressor oil pack	COpack

Codes for finished lubricants should be distinct from those for additives or packages, for example:

Engine oils	EO
Gear oils	GO
Automatic transmission fluids	AT
Hydraulic fluids	HF
Compressor oils	CO
Turbine oils	TO
Machinery oils	MO
Metalworking fluids	MW

The result of this coding can be used to maintain the confidentiality of formulations. For example, the formulation for a hydraulic oil might appear on the automated blending system as

HF-001	
Component	% wt
BO-02	96.95
AO-04	1.00
AW-06	1.00
RI-03	0.50
MP-11	0.05
PP-08	0.50

Access to the codes should only be allowed for authorised staff, who have password-protected access to both the computer system and the product formulary. Great care should also be taken to ensure that all incoming raw materials are correctly coded and only by those authorised staff who have access to the codes.

11.4.6 Batch Numbering and Traceability

Each batch of blended lubricant should be assigned a unique batch number, preferably with some type of date code. This batch number must then be used on all packages and for all bulk deliveries of that finished lubricant, as well as for transportation documentation.

Modern automated blending systems are programmed to assign a unique batch number to each blend. Also, mixtures of batches must be given a unique batch number, again with a code for the date of mixing.

The primary reason for assigning unique batch numbers is to enable the traceability of each product, from raw material delivery, through storage, blending, filling and packaging, warehousing and delivery. Traceability is very important as part of resolving future customer complaints, as discussed below. If a customer experiences a problem with a specific batch of lubricant (or even with a mixture of batches), it will be possible to determine when and how the batch was made, which raw materials were used and how the product was packaged, stored and delivered.

11.4.7 Computerised Blend Records

Again, modern automated blending systems are programmed to record and store information pertaining to every blend. Details of raw materials, formulations, weights of components, blending equipment, blending conditions, testing and test results, packaging and packing conditions and storage conditions can be stored easily on the distributed control system (DCS) server.

As with raw material and formulation codes, access to these records (whether for retrieval or updating) should only be possible for authorised staff, in order to preclude any tampering with potentially vital information.

11.4.8 Blend Sample Storage and Retention

Samples of every blend, taken as part of the testing of finished products, should be retained and stored in a secure location. At least 1 L of each batch should be stored, although some products may need to have 2, or even 5 L of sample retained.

The sample store should be large enough to contain several thousand samples at room temperature (around 20°C) and in the dark. An ideal location for the sample store is immediately adjacent to the blending plant's test laboratory, where only laboratory and supervisory staff have access to the store.

Blend samples should be retained for at least 2 years, and preferably 3 years. This will ensure that sufficient time has elapsed between delivery of a lubricant to a customer and the resolution of any future customer complaint.

11.5 RESOLVING PRODUCT QUALITY PROBLEMS

Complaints can arise internally within a company that uses lubricants, for example, that a machine is not behaving well with the lubricant that has been supplied. Or they can be external complaints by a customer to the supplier of the lubricant. While attitudes and approaches may differ somewhat, the basic rules for dealing with complaints are essentially the same in both cases. For maximum coverage, we will treat the subject as though it is the latter form of complaint and consider some ground rules as to how the lubricant supplier can handle it.

A policy for handling complaints should be developed and communicated to all personnel involved. Rapid action, replacement of material and compensation for damage can often prevent the loss of a customer who has complained. However, too-ready acceptance of responsibility for a problem can become expensive, particularly if compensation for mechanical damage and loss of business are concerned. A compromise is to immediately replace any lubricant which is suspected of being either contaminated or off grade, but to resist suggestions that mechanical damage is directly associated with lubricant quality until an exhaustive investigation has taken place. Field workers investigating complaints must know precisely what offers they can make to an aggrieved customer.

It is most important that complaints be properly documented, and it is well worthwhile to develop special forms for the purpose. Details of oil grade, delivery date, batch number and customer storage conditions need to be ascertained, plus information on the equipment type, duty and age or condition. A sample of the oil which is the subject of a complaint needs to be obtained, and a sample of used oil if the complaint concerns used oil condition or equipment malfunction. The taking of samples requires particular care and attention. A representative average sample must be obtained, or specific top, middle and bottom samples in the case of non-homogeneous material. A copy of the equipment manufacturer's recommendations should also be obtained, or their contact address and telephone number.

It is usually clear by this time whether the complaint is relatively trivial or if it has potentially serious implications. In the case of contaminated unused oil, the only testing necessary may be a visual examination without specific identification of any contamination or sediments. If the sample appears normal, it should be compared with the retained sample which was taken at the blending plant before delivery. If they appear identical, correctness of grade can be further confirmed by tests such as base number, sulphated ash or metals analysis. If desired, contamination can be checked for by flash point, water content and filtration of the sample.

If the unused oil appears to be within specification, and the complaint concerns its performance in a machine, the first action it to check whether the grade is one recommended for that machine and the service conditions. The laboratory should examine used oil samples for contamination, fuel diluent, insolubles, acid number and base number, and elemental contents. From these results, it will be possible to assess the overall condition of the oil, the mechanical condition of the machine and/or the length of time the oil has been in service. A judgement will then have to be made based on experience as to whether the problem is one of oil quality, machine condition or the service conditions.

The usual causes of complaints are:

- Wrong grade of oil supplied or used: Easily detected in the laboratory, if not from package or drum labels.
- Contamination: Usually detectable by appearance or smell; can take place at the blending plant, in transit (bulk) or at the customer's premises.
- Wrong oil recommendation: To be assessed by a sales engineer after considering engine or equipment type and service.
- Poor engine or equipment condition: Can be assessed by examination of the used oil for fuel dilution, sludge, wear metals and other factors.

- Extreme service conditions: Detectable from insoluble contents, viscosity, wear metals and so on, provided that maintenance procedures are known. The user may also provide information on the type and severity of service.
- Poor maintenance: Can produce symptoms similar to either or both of the above cases, and has to be assessed more from a knowledge of the customer's operations.
- Mechanical failure: Often the basic origin of the complaint, and does not need specific confirmation. However, the parts should be examined as carefully as possible by a trained technical service engineer to try to determine a sequence of events.

Complaints of mechanical failure, blaming the detergency of the lubricant as the primary cause, are frequently unjustified. An engine maintained in good condition and operated within its normal duties should operate successfully without problems on the specified quality of lubricant. International standards of lubricant quality contain sufficient safety margins that normal-quality variations will produce no detrimental effects. (An exception was the specification of low-temperature flow qualities after standing at very low temperatures. New tests on cold cranking and low-temperature flow had to be developed after a series of engine failures due to low-temperature operability problems.)

A common cause of problems is poor maintenance, which can lead to either filter blocking or, in other cases, excessive fuel dilution and consequent loss of oil viscosity. Diesel injector maintenance is particularly important for a heavy-duty diesel engine, and in the past many mechanical problems blamed on the lubricant have turned out to be caused by dribbling injectors producing excessive fuel dilution.

Complaints of short overhaul lifetimes due to poor lubricant performance are harder to deal with. Assuming that a maker's recommended quality of lubricant is in use, either the equipment is old and/or poorly maintained or the service conditions are more severe than the user realises. (It is a common misapprehension that engines spending much time idling stress lubricant less than when working hard. The opposite may often be the case.) One solution here may be to provide a higher-quality lubricant for a trial period at no increased cost, and to monitor the effect with the user's cooperation.

It is worth noting that, in the author's experience, approximately three-quarters of all customers' complaints about lubricants have not been caused by poor-quality lubricants.

11.6 SUMMARY

Controlling quality during the blending of lubricants is relatively straightforward, once the correct procedures and methodologies have been planned and established as part of a total QMS.

Since quality is defined by customers, all that a blending plant needs to do is to ensure that each blend of each product meets the specification or requirements of customers, as established by the marketing and sales team.

Each function in the blending plant needs to participate in the quality control process. Fortunately, with lubricant blending, automation and simple, quick tests help operators, technicians and managers considerably.

12 Lubricant Packaging and Filling

12.1 INTRODUCTION

Lubricants are delivered to customers either in bulk or packaged in containers of many different types. In every stage of packaging, handling, stacking, storing and dispensing lubricants, benefits are obtained from observing good housekeeping and using recommended techniques.

Some methods of handling and storing lubricants are essential for safety reasons. Others will ensure that the lubricants do not become contaminated before use. Others relate to avoiding or minimising damage to the environment.

The care with which lubricants are formulated and blended can be completely nullified by incorrect packaging, unsatisfactory storage or careless handling. Damage to packages can lead to leakage, which may cause an unsafe workplace or environmental pollution, contamination of the contents or obscuring of labels. Unsatisfactory bulk storage can also lead to contamination of the lubricant or confusion between grades. The wrong product could be used in the machines to be lubricated.

This chapter aims to describe and review the main methods, and their benefits, for packaging and handling lubricants. This will help to ensure that the right lubricant, free from contamination, reaches the point of application. It will also help to avert unsafe working conditions, fire hazards and pollution. Overall, correct lubricant packaging, handling and storage will help achieve maximum operating efficiency by blending plants, for the ultimate benefit of customers.

12.2 ROLE AND ATTRIBUTES OF PACKAGING

Packaging can serve a number of purposes:

- To protect the product and maintain its quality
- To stop the product impacting other products and the environment
- To provide units of the product that are easily manageable by users
- To help promote the product

In retail consumer marketing particularly, the promotional aspect aims to make the product stand out on a store shelf and say “buy me” to the customer walking down the store aisle. Many retail packages, especially perishable foods, expensive products and drugs, must be tamper-proof to the extent that the consumer can determine whether the package contents have been altered or contaminated. The choice of packaging materials also is influenced by concerns for environmental protection.

Containers that can be recycled, or are made of recycled materials, are being used increasingly.

Packaging can play an important role in reducing the security risks of transportation, handling, display and selling. Packages can be made with improved tamper resistance, to deter manipulation, and they can also have tamper-evident features that indicate that tampering has occurred. Packages can be engineered to help reduce the risks of package pilferage or the theft and resale of products. Some package fabrication is more resistant to pilferage than other types, and some have pilfer-indicating seals. Counterfeit consumer goods, unauthorised sales (diversion), material substitution and tampering can all be minimised or prevented with such anti-counterfeiting methods. Packages may include authentication seals and use security printing to help indicate that the package and contents are not counterfeit. Packages also can include anti-theft devices such as dye packs, radio frequency identification (RFID) tags or electronic article surveillance tags that can be activated or detected by devices at sales or retail exit points and require specialised tools to deactivate. Using packaging in this way is a means of retail loss prevention.

Most retail products are packed in a hierarchy of packaging. The concept can be compared with building blocks. The smallest size is the container that the customer buys and takes home. These containers fit into boxes that are about 1 cubic foot (0.028 m^3) in dimension and are unloaded, item by item, by the person stocking the shelves. These boxes in turn are handled on pallets, wooden or recycled plastic platforms about 6 inches high and 40 by 48 inches along the top ($0.15 \times 1.02 \times 1.22$ m). Pallets are loaded two or four boxes high and moved by forklift trucks into and out of warehouses, rail wagons and trucks. Pallet loads, also called "unit loads", are the most common way of handling packaged freight.

Products that are not packaged are often handled in bulk. Iron ore, coal and grains are moved in truckload, trainload and shipload lots. They are loaded, unloaded and transferred by large mechanical devices. Liquids such as crude oil, petroleum products and petrochemicals are pumped through pipelines or carried in tankers. Flour and cement are moved between dry tanks pneumatically.

While the origins of packaging can be traced to the leather, glass and clay containers of the earliest Western cultures, its economic significance has increased dramatically since the start of the Industrial Revolution. Packages must also be easy to manufacture and fill. They must also be inexpensive relative to the price of the final packaged product.

Materials used in the transport of substantial loads of goods include corrugated or solid cardboard for lighter materials, metal or plastic for liquids and wood for heavy or bulky loads. Timber cases and crates are widely used for weights of more than 100 kg, while below this weight fibreboard, either solid or corrugated, is the favoured material. Wooden pallets have replaced crates in some instances. Plastic has also been used extensively as an impact buffer and, because of its high durability and insulation qualities, as a shipping material for liquids and perishable foods. The most widely used material in the packaging of goods for consumers is the cardboard carton, since it is light in weight, is inexpensive and can be manufactured, printed and stored easily.

Cans of tin-plated steel, both those that are permanently sealed and those with tops that can be lifted and replaced, are also used predominantly for food storage.

Tin-plate containers are also used to hold paints and varnishes and tobacco, medical and cosmetic products. While tin plate is durable and highly resistant to chemical and mechanical damage, aluminium is lighter and more malleable, but interacts more readily with chemical agents. Aluminium is used for beverage cans and provides bottle caps and easy-open tops for other cans. Most aerosol (pressurised) containers, which deliver liquid products in the form of a spray, are based on metal cans.

Glass containers are easily mass-produced, can be reused, are durable, are resistant to almost all chemicals and can be kept highly sanitary. They are therefore ideal for the storage of solid and liquid foods, drugs and cosmetic products. Polyvinyl chloride (PVC), polypropylene (PP), polyethylene and polystyrene (PS) plastics are used as packaging materials. Plastics are often produced in the form of trays, bags, bottles, boxes and transparent film, through thermoforming and injection moulding or blow moulding processes. Their light weight, flexibility and insulating qualities make them especially useful for pressurised packages and containers of foods to be boiled or frozen. Collapsible plastic tubes are widely used to hold cosmetics, toiletries and pharmaceuticals.

Labelling for packages must be easy to print and to affix to the container material.

12.3 TYPES OF LUBRICANT PACKAGING

12.3.1 Plastic Bottles

Plastic bottles, made from high-density polyethylene (HDPE), PP or polyethylene terephthalate (PET), offer the advantage of breakage resistance and lightness, so they are ideal for distributing and storing small quantities of lubricants. Standard sizes are 5, 4, 2, 1 and ½ L in Europe and many other regions and 1 gallon, 1 quart and 1 pint in North America. Most plastic bottles for oils are coloured and opaque. The colouring is used to assist brand and product recognition in retail markets. Four- and five-litre plastic oil bottles are either designed for easy pour or have easy-pour spout inserts. All plastic oil bottles have tamper-proof plastic caps. Examples of several designs of plastic bottles for lubricants are shown in Figure 12.1.

The disposal of plastic bottles contributes to pollution, because few plastic containers disintegrate upon exposure to the environment. Plastic recycling (discussed later) was instituted in the early 1990s to help reduce the solid-waste problem. Most plastic oil bottles in Europe and North America are now made of trilayered blow-moulded high-density plastic, with a layer of recycled plastic sandwiched between an inner layer and an outer layer of virgin plastic. The inner layer protects the oil, while the outer layer protects the environment and the consumer, and can still be screen-printed or have a label attached to it. In this way, between 40% and 50% of the plastic bottle can be recycled.

In the last few years, some North American lubricant marketers have introduced clear PET bottles for lubricants, so that retail customers can see the light colour of the contents. Light oil colour is being associated increasingly with synthetic and high-performance, and hence higher-added-value, lubricants.

Until the mid-1970s, small quantities of lubricants were sold in 1-gallon (4.5 L), 1-quart (1.1 L) or 1-pint (568 ml) tin-plate cans. The 1-gallon cans were of a

FIGURE 12.1 Examples of automotive engine oil bottles.

rectangular section, with the handle and the hole both at the top. A few manufacturers have reverted to supplying specialist automotive oils in these tin-plate cans, for marketing image and branding reasons.

Also, from the early 1950s to the late 1980s, lubricants were stored and dispensed on gasoline service station forecourts in clear glass bottles with screw-on tin-plate pouring spouts. Bottles that had been used to top up motorists' oil during the day were refilled every evening from a drum of the correct oil stored in the service station workshop. Glass bottles afford highly effective protection of their contents and are attractive because of their transparency and high gloss and the variety of shapes attainable. Fragility is a major disadvantage, and only coloured glass protects those products sensitive to the action of light.

Returnable glass bottles, which can be reused a number of times, are the least expensive to manufacture on a per use basis, although repeated handling costs may reduce any savings. It is possible that full-service oil top-up on service station forecourts may be reintroduced at some future time, to help reduce the problem of waste packaging collection and recycling.

12.3.2 Stand-Up Pouches

A recent innovation with packaging lubricants, introduced in the United States, Mexico and a number of countries in Asia, is the flexible plastic stand-up pouch. Stand-up pouches can incorporate convenient features, such as spouts and handles to facilitate easy pouring. Pouches also provide the benefit of much lighter package weights, since pouches often use approximately 60% less plastic than conventional rigid plastic bottles. A photograph of a stand-up pouch is shown in Figure 12.2.

FIGURE 12.2 Photograph of a stand-up pouch. (From Lube-Tech and Glenroy Inc. With permission.)

The suppliers of stand-up pouches promote several advantages for them in comparison with rigid plastic bottles. They claim that a plastic stand-up pouch provides approximately seven times more printed surface area on which to display marketing and branding graphics. The claimed environmental benefits include lower energy requirements during production (meaning lower production of carbon dioxide); fewer trucks for transporting empty packages, again leading to lower energy use; and a higher product-to-package ratio. This produces significantly less landfill waste through source reduction, although this does not take into account plastic recycling. Stand-up plastic pouches occupy less shelf space than rigid plastic bottles in retail lubricant marketing channels.

The manufacturers of stand-up plastic pouches engineer them to be extremely durable, with claimed excellent seal strength in order to withstand rigorous burst testing and drop testing. The plastics used are resistant to oil migration and delamination.

However, at the time of writing there are very few suppliers of stand-up plastic pouches and lubricant blending plants are required to purchase them in large quantities, empty and preprinted with marketing and branding graphics. For a lubricant blending plant with its own blow moulding equipment for producing rigid plastic bottles on site, the economics of using plastic stand-up pouches may not be cost-effective at present.

12.3.3 Oil Drums

An oil drum, which has a capacity of 205 L (45 UK [imperial] gallons, 54 US gallons), is the largest container for packed lubricants. They were usually made of 18-gauge (1.2 mm or 0.048-inch) steel, as used for industrial deliveries. They are about 890 mm high and 600 mm diameter (35 × 23 inches) and have two threaded bungholes at one end, 50 and 20 mm in diameter (2.0 and 0.75 inches, respectively). Early drums also had a bunghole in the sidewall.

The term *drum* is now being used for oil products, so that it is not confused with *barrel*, the oil industry's international measure of crude oil output or refinery throughput. A barrel is a quantity equal to 35 UK gallons or 42 U.S. gallons.

A number of manufacturers now supply drums made of HDPE. They have the same capacity as a steel drum and may have the same (or very similar) shape and dimensions. Some plastic drums are shaped more like wine or beer barrels. They are easier and cheaper to make and recondition than are steel drums.

While plastic drums are more robust to mishandling (denting) than steel drums, they are less resilient to being dropped from heights above about 4 feet when full. The main disadvantages of plastic drums are that they cannot be stacked unsupported more than three high (steel drums can be stacked unsupported five or six high, although this is not recommended for safety reasons). Also, they tend to become misshapen when stored outdoors in hot climates.

Stacking plastic drums on pallets is also a problem if there is a variation in their height of more than about 0.5 cm (¼ inch). Plastic drums tend to slide off pallets when forklift trucks are manoeuvring in delivery and storage areas. Many pallet manufacturers now supply small triangular corner boards to prevent this happening. Also, plastic drums cannot be used to store viscous lubricants, because they cannot be heated in an oven to allow the contents to be poured or extracted more easily. Examples of different types of drums are shown in Figure 12.3.

Reconditioning and reuse of plastic drums is more cost-effective than steel drums. On average, steel drums are reused between four and six times. In comparison, plastic drums can last two or three times longer than a steel drum.

FIGURE 12.3 Examples of drums and IBCs.

12.3.4 50 L (10-Gallon) Drums

This size of drum is now used only rarely. It is 500 mm high and 380 mm in diameter (20 × 15 inches) and holds 45 L.

12.3.5 25 L (5-Gallon) Drums

These are 430 mm high and 300 mm in diameter (17 × 12 inches), have an integral handle at the top and hold 25 L. Like 205 L drums, they have two bungholes at the top, 50 and 20 mm in diameter.

Twenty- or twenty-five-litre plastic drums are now available, as are 120 and 150 L plastic drums. Both are made from HDPE using blow moulding methods and are used for detergents, pharmaceuticals and agrochemicals, as well as for lubricants.

12.3.6 Grease Drums, Kegs, Pails and Cans

A grease drum has approximately the same dimensions as an oil drum and holds 180 kg (400 lb) of grease. Its flat lid is secured by a circumferential channel-section strap.

A grease keg is 610 mm high and 360 mm diameter (24 × 14 inches) and holds 50 kg (112 lb) of grease. The open top is covered by a flat lid held in place by lugs.

A grease pail is 330 mm high and 250 mm diameter (13 × 10 inches), and the open top has a flat, lugged lid. Pails hold 25 kg of grease and are more common than kegs because they are easier to handle.

Smaller quantities of grease are delivered in tin-plate or plastic cans which hold 3 kg, 1 kg or 500 g of grease. These packages are useful for retail consumers and smaller commercial and industrial users of grease.

Increasingly in the commercial and industrial markets, greases are being packed and delivered in special delivery tubes that are connected directly to the equipment to be lubricated. These tubes minimise the risks associated with storing and dispensing greases.

12.3.7 Intermediate Bulk Containers

Many manufacturers of plastic drums now supply intermediate bulk containers (IBCs) with capacities ranging from 250 to 1000 L. Most are rectangular and are either designed to stand on a pallet or have an integrated "forkliftable" base. Many are produced through an advanced blow moulding process with co-extrusion of two types of HDPE. The outer layer provides the required strength and protection from UV radiation, while the inner layer is resistant to most chemicals. The floors of IBCs are generally dome shaped, to provide optimum drainage.

Some IBCs are contained within a tubular thermogalvanised steel frame, which is connected to the pallet base to provide enhanced sturdiness and protection. Another design of IBC features an inner bag, which is replaced after each use. This type avoids the need for cleaning before reuse. It also combines guaranteed product integrity with the lowest cost per trip.

An even later design of IBC features a collapsible outer corrugated board container with an inner liner bag. This is particularly suitable for transporting and storing non-hazardous viscous liquids. These IBCs are supplied either flat packed or built up. The flat pack saves costs when transporting empty containers. Although they are designed as one-trip containers, they can be reused (with a new liner bag) if treated carefully and the materials are recyclable. Examples of IBCs are shown in Figure 12.3, together with examples of drums.

12.3.8 Additional Packaging

Smaller containers need additional protection and are packed in corrugated cardboard boxes or fibreboard cases. Standard groupings are:

- 4×4 or 5 L.
- 6×2 L.
- 12×1 L or quart.
- 12 × ½ L.
- 24 × ½ L.

The cases are palletised, and some users stack the pallets in an interlocking pattern (like brickwork) for stability and extra compressive strength. While pallets enable safe stacking to reasonable heights, usually up to four layers, steel racks are sometimes more convenient since they permit packages to be grouped ready for dispatch.

12.4 RECONDITIONING DRUMS AND IBCS

Steel drums are sturdy enough to withstand several trips, but they may then be in poor condition. Cleaning and reconditioning techniques have been developed, and workshops for this purpose are located close to larger blending plants. The process for reconditioning steel drums involves thorough cleaning inside and out, removal of all dents and rerolling the seams. Then the drums are tested for leaks, and finally they are rinsed and dried. Before they are repainted, their interiors are inspected optically and any residual material is removed by vacuum. The bungs are then screwed home and remain in place until immediately before the filling operation is due to begin. Drums are usually given a final inspection in the filling area.

Reconditioning plastic drums and IBCs is somewhat similar. Old labels are removed and the containers are cleaned and dried, inside and out. Detergents or solvent cleaners are used for cleaning, with facilities for their recovery and reuse. Mild solvents are usually used for rinsing, before drying. Again, the cleaned drums or IBCs are pressure tested to check for leaks, their interiors are inspected optically and any residual materials are removed. Bungs are replaced, but generally not tightened.

Twenty-five- litre drums and 5 , 2 and 1 L plastic bottles are used only once. (Some users of lubricants reuse 25 L drums, but usually only for handling and dispensing smaller quantities of oil taken from 205 L drums or bulk storage tanks.) Drums are received from the makers with the bungs and caps in place. They are inspected internally before being filled. Plastic bottles, which are used in huge numbers, are supplied with their closures separate.

12.5 RECYCLING PLASTIC PACKAGING

Sanitary landfills, in which layers of refuse alternate with layers of soil, are the favoured method of disposing of solid waste in many regions and countries. However, concerns over the sustainability of such land use has encouraged efforts to dispose of various materials by recycling them for reuse or to derive some positive benefits. Paper, glass and aluminium containers have been recycled to an increasing extent for many years. Recycling plastic packaging has become more common.

Several technical and economic problems are encountered in the recycling of plastics. They fall into two general categories:

- **Identification, segregation (or sorting) and gathering into central locations.** Since plastics used in packaging form a highly visible part (approximately 20% by volume but less than 10% by weight) of general wastes, most recycling efforts have focused on containers. Almost all bottles, food trays, cups and dishes made of the major commodity plastics now bear an identifying number enclosed in a triangle together with an abbreviation. This makes the plastics easier to sort, although it can still be labour-intensive. In addition to such labelling, some consumers are encouraged to return empty beverage containers to the place of purchase by being required to pay a deposit on each unit at the time of purchase. This system helps to solve two of the major problems associated with economical recycling, since the consumer seeking return of the deposit does the sorting and the retailers gather the plastics into central locations. An added attraction of deposit laws is a notable decrease in environmental litter.
- **Economics of recovering value.** Although thermoplastics can be recycled more readily than thermoset plastics, there are inherent limitations on the recycling of even these materials. First, a recyclable plastic may be contaminated by non-plastics or by different polymers making up the original product. Even within a single polymer type, there are differences in molecular weight. For instance, a supplier of PS may produce a material of high molecular weight for sheet-formed food trays, since that forming process favours a high melt viscosity and elasticity. At the same time, the supplier may offer a low-molecular-weight PS for the injection moulding of disposable dinnerware, since injection moulding works best with a melt of low viscosity and very little elasticity. If the polymers from both types of product are mixed in a recycling operation, the mixed material will not be very suitable for either of the original applications.

Another complication to the recycling of plastics is the mixing together of pigments or dyes of different colours. Yet another is the problem of quality control. Almost all plastics change either slightly or greatly as a result of initial fabrication and use. Some, for instance, undergo changes in molecular weight due to cross-linking or chain scission (breaking of the chemical bonds that hold a polymer chain together). Others undergo oxidation, another common reaction that can also change the properties of a plastic.

For all the foregoing reasons, recycled plastics will almost always have certain disadvantages in comparison with virgin plastics. Most thermoplastics are therefore recycled into somewhat less demanding applications. HDPE from thin-walled grocery bags, for example, may be converted into thick-walled flowerpots, PVC recovered from bottles may be used in traffic cones and PET recovered from beverage bottles may be washed, dried and melt-spun into fibrous filling for pillows and clothing.

Waste plastics that cannot be separated by polymer type can be made into plastic "timber", extruded slabs that are suitable for applications such as industrial flooring, pallets and park benches. The heterogeneous composition of plastic timber makes it inherently weaker than the original polymers. Other recycling processes that make use of mixed plastics are pyrolysis, which converts the solids into a petroleum-like substance, and direct incineration, which can provide energy for power plants or industrial furnaces.

In most plastic recycling operations, the first step after sorting is to chop and grind the plastic into chips, which are easier to clean and handle in subsequent steps. The chips commonly are first washed in order to remove non-plastic items, such as labels, caps and adhesives. If the material comes from a narrowly defined source, it may be possible to dry the washed chips and immediately extrude them into moulding pellets or even to extrude them directly into fibres. For "mixed-waste" polymers, automatic separation processes based on differences in density or solubility have been used to some extent.

In Europe, packaging and packaging waste are regulated under the European Commission (EC) Packaging Directive (94/62/EC), under which all member states have agreed to meet specific standards and targets. As a result, all European countries have now instituted national regulations that include standards for the type and quality of packaging and targets for the recovery and recycling of packaging waste.

The regulations place obligations on certain businesses to recover and recycle specified tonnages of packaging waste, based on the amount of packaging handled by the business. For example, by 2001, at least 50% of packaging waste in the United Kingdom was being recovered. Any business handling more than 50 tonnes of packaging annually and with a turnover of more than £2 million was obligated if it performs one or more of the following activities:

- Manufacturing raw materials for packaging.
- Converting materials into packaging.
- Filling packaging.
- Selling filled packaging.
- Importing packaging or packaging materials.

Currently, the activity obligations set out in the regulations are raw material manufacturing 6%, converting 9%, pack/filling 37% and selling 48%.

The directive was amended in 2004 to provide criteria that clarified the definition of the term *packaging* and to increase the targets for recovery and recycling of packaging waste. It was revised again in 2005 to grant new member states transitional periods for attaining the recovery and recycling targets. In 2013, Annex I of the directive, containing the list of illustrative examples of items that are or are not to be considered as packaging, was revised in order to provide more clarity by adding a number of examples to the list. The latest revision of the Packaging and Packaging Waste Directive occurred in April 2015 with the adoption of Directive (EU) 2015/720 of the European Parliament and of the council-amending Directive 94/62/EC as regards the consumption of lightweight plastic carrier bags.

To illustrate, the United Kingdom has the Producer Responsibility Obligations (Packaging Waste) Regulations 1997 and the Packaging Waste Recovery Note (PRN) system. Each company covered by the regulations has a legal obligation to cause the recovery of a calculated tonnage of waste packaging. This can be achieved either by setting up its own arrangements or by joining a collective compliance scheme that will undertake recovery on behalf of its members. PRNs are the principal form of documented evidence that materials have been recovered. They are unique numbered documents provided by the environment agencies to accredited reprocessors.

Targets are in place in each country in Europe for the recovery and recycling of each type of packaging material, including steel and plastics. Each European country has a huge range of plans and activities aimed at meeting these targets. Germany, for example, was significantly ahead of France, Italy and the United Kingdom by 2012.

At the end of 2017, the EC announced plans to make all plastic packaging across Europe recyclable, reusable or degradable by 2030. This forms part of a wider strategy to tackle the issue of plastic waste. The strategy includes commitments to reduce the consumption of single-use plastics and restrict the use of microplastics, such as microbeads found in some cosmetics. At present, Europeans produce 25 million tonnes of plastic waste every year, but less than 30% of this is recycled. To aid the increase in recycling rates, the EC will provide 100 million Euros to finance the development of "smarter and more recyclable plastics materials".

The EC plan appears to be a response to China informing the World Trade Organization in July 2017 that it would no longer accept imports of "foreign garbage" from January 2018. The ban affects waste paper and plastics and is aimed at encouraging recycling within China and helping to clean up its industrial pollution. One immediate effect of the ban has been that China is now importing more virgin plastic, mainly from the United States. Another effect is that other countries in Asia have begun to set up packaging recycling facilities.

In California, the mandatory plastics recycling rate of 25% was raised to 30% in January 2001 and to 35% in January 2003. Around 12 years ago, several major U.S. oil companies attempted to implement engine oil bottle recycling programmes. Chevron set up an oil bottle recycling programme on the northwest coast to collect its own bottles and recycle them into new Chevron bottles. While the programme was a technical success, it was not economically. Amoco had a pilot programme in Chicago in the early 1990s, Exxon experimented with oil bottle collection and Sun had a pilot bottle recycling programme in Philadelphia, but none of these programmes succeeded.

At the end of 1999, North America had three engine oil bottle recycling programmes. A grant from the federal Environmental Protection Agency was used to start a pilot programme in two South Carolina counties in 1996. The oil bottles were collected from homes and drop-off centres, separated, ground into flake and sold to KW Plastics in Troy, Alabama. The second programme, project GEORGE, was operated by Fixcor with a grant from California; AlliedSignal owns the patent on the process. The project collected the oil residue from the bottles and processed the plastic without the use of water, chemicals or detergents. While initial testing in California has been completed, the project needed financial support to continue. The Alberta Used Oil Management Association (AUOMA), in Alberta, Canada, was the third programme, funded by an industry-operated recycling fund generated by a handling charge on lubricating oil materials. AUOMA collects used oil, filters and bottles, charging 5 cents (Canadian) per litre or kilo for waste oil, 5 cents per litre for containers and 50 cents or $1 per filter. This programme is succeeding and expanding across Alberta.

The main observation from all the programmes is that plastic oil bottle recycling as a stand-alone process is not economically sustainable. Collection is the primary difficulty, waste oil in the bottles presents special problems and not all HDPE recycling companies that can recycle coloured bottles can recycle oil bottles. It was found that used plastic oil bottles, even when they have been drained thoroughly, still contain small amounts of oil, which makes recycling them particularly difficult. In almost all cases, it is more practical and economic to incinerate used plastic oil bottles, to recover the thermal value of the plastic and the oil.

12.6 LABELS FOR LUBRICANT PACKAGES

12.6.1 Role of Labels on Packages

Labels can be used to communicate how to use, transport, handle, recycle and/or dispose of the package or product. Labels can also be used for track and trace purposes, particularly if there are batch numbers on the label. Labels on packages may indicate the package's construction material, particularly for plastic packages, with a symbol.

Labels can also be used by marketers as part of the promotion of a product, to encourage potential buyers to purchase the product. Package graphic design and physical design have been important and constantly evolving phenomena for several decades. Marketing communications and graphic design are applied to the surface of the package and often to the point-of-sale display. Most packaging is designed to reflect the brand's message and identity.

There are also legal requirements for labels, to comply with international legislation, national legislation and consumer protection regulations. Labels may display health and safety information, as well as used product and package collection and recycling. For lubricants, labels will usually indicate what to do in the event of a spillage. There may also be methods for disposal of used oil. It is worth noting that new oils are generally non-hazardous, while in many countries, used oils are usually classified as hazardous. (A discussion of this is outside the scope of this book.)

12.6.2 The Globally Harmonised System

The Globally Harmonized System of Classification and Labelling of Chemicals (GHS) was created by the United Nations (UN) to replace several classification and labelling standards used in different countries. The GHS uses consistent criteria for classification and labelling on a global level. Its development began at the UN Rio Conference in 1992. It supersedes the relevant European Union (EU) system (which has implemented the GHS into EU law as the CLP regulation) and U.S. standards.

Other regulations for the classification and labelling of goods include ADR (Canadian Transportation of Dangerous Goods), DOT (U.S. Department of Transport), IMDG (International Maritime Dangerous Goods Code) and IATA (International Air Transport Association Dangerous Goods Regulations), all of which have been replaced by the GHS, but all of which also include regulations and guidelines for the packaging and transportation of these goods. (Transportation of base oils, additives and lubricants is outside the scope of this book.)

The European regulation on Classification, Labelling and Packaging of Substances and Mixtures (CLP) redefines the classification criteria for the physical, toxicological and environmental properties of substances and mixtures and harmonises hazards-related communication.

Before the GHS was devised and implemented, there were many different regulations on hazard classification of chemicals used in different countries. While those systems may have been similar, they resulted in multiple standards and classifications and labels for the same hazard in different countries. In view of the extent of international trade in chemicals and the potential impact on countries when controls are not implemented, it was determined that a worldwide approach was necessary.

The GHS is not compulsory under UN law, but provides the infrastructure for participating countries to implement a hazard classification and communication system, which many less economically developed countries would not have had the money to create themselves. In the longer term, the GHS is expected to improve knowledge of the chronic health hazards of chemicals and encourage a move toward the elimination of hazardous chemicals, especially carcinogens, mutagens and reproductive toxins, or their replacement with less hazardous ones.

The GHS classification system is a complex system with data obtained from tests, literature and practical experience. The three elements of the hazard classification criteria are physical, health and environmental.

Physical hazards are largely based on those of the UN Dangerous Goods System, although some additions and changes were necessary since the scope of the GHS includes all target audiences. The physical hazards are:

- Explosives.
- Flammable gases.
- Flammable aerosols.
- Oxidising gases.
- Gases under pressure.
- Flammable liquids.
- Flammable solids.

- Self-reactive substances (thermally unstable liquids or solids).
- Pyrophoric liquids.
- Pyrophoric solids.
- Self-heating substances.
- Substances which on contact with water emit flammable gases.
- Oxidising liquids.
- Oxidising solids.
- Organic peroxides.
- Substances corrosive to metal.

In the GHS, health hazards are:

- Acute toxicity.
- Skin corrosion.
- Skin irritation.
- Serious eye damage.
- Eye irritation.
- Respiratory sensitiser.
- Skin sensitiser.
- Germ cell mutagenicity.
- Carcinogenicity.
- Reproductive toxicity.
- Specific target organ toxicity (STOT).
- Aspiration hazard.

Environmental hazards are:

- Acute aquatic toxicity.
- Chronic aquatic toxicity.

The GHS approach to the classification of mixtures for health and environmental hazards is also complex. It uses a tiered approach and is dependent on the amount of information available for the mixture itself and for its components. Principles have been developed for the classification of mixtures, drawing on existing systems such as the EU system for classification of preparations laid down in Directive 1999/45/EC. The process for the classification of mixtures is based on the following steps:

- Where toxicological or ecotoxicological test data are available for the mixture itself, the classification of the mixture will be based on that data.
- Where test data are not available for the mixture itself, the appropriate bridging principles should be applied, which use test data for components and/or similar mixtures.
- If test data are not available for the mixture itself, and the bridging principles cannot be applied, then use the calculation or cutoff values described in the specific endpoint to classify the mixture.

The GHS document does not include testing requirements for substances or mixtures. In fact, one of the main goals of the GHS is to reduce the need for animal testing. The GHS criteria for determining health and environmental hazards are test method neutral, allowing different approaches as long as they are scientifically sound and validated according to international procedures and criteria already referred to in existing systems. Test data already generated for the classification of chemicals under existing systems should be accepted when classifying these chemicals under the GHS, thereby avoiding duplicative testing and the unnecessary use of test animals. The GHS physical hazard criteria are linked to specific UN test methods. It is assumed that mixtures will be tested for physical hazards.

After the substance or mixture has been classified according to the GHS criteria, the hazards need to be communicated. As with many existing systems, the communication methods incorporated in GHS include labels and safety data sheets (SDSs). The GHS attempts to standardise hazard communication so that the intended audience can better understand the hazards of the chemicals in use. The GHS has established guiding principles:

- The problem of trade secret or confidential business information has not been addressed within the GHS, except in general terms. For example, non-disclosure of confidential business information should not compromise the health and safety of users.
- Hazard communication should be available in more than one form (for example, placards, labels or SDSs).
- Hazard communication should include hazard statements and precautionary statements.
- Hazard communication information should be easy to understand and standardised.
- Hazard communication phrases should be consistent with each other to reduce confusion.
- Hazard communication should take into account all existing research and any new evidence.

Comprehensibility is challenging for a single culture and language. Global harmonisation has numerous complexities. Some factors that affected the work included:

- Different philosophies in existing systems on how and what should be communicated.
- Language differences around the world.
- Ability to translate phrases meaningfully.
- Ability to understand and appropriately respond to symbols or pictograms.

These factors were considered in developing the GHS communication tools. The standardised label elements included in the GHS are:

- **Symbols (GHS hazard pictograms):** Convey health, physical and environmental hazard information, assigned to a GHS hazard class and category. Pictograms include the harmonised hazard symbols plus other graphic elements, such as borders, background patterns or cozers and substances

which have target organ toxicity. Pictograms will have a black symbol on a white background with a red diamond frame. For transport, pictograms will have the background, symbol and colours currently used in the "UN Recommendations on the Transport of Dangerous Goods". Where a transport pictogram appears, the GHS pictogram for the same hazard should not appear. The pictograms and downloadable files can be accessed on the UN website for the GHS at
http://www.unece.org/trans/danger/publi/ghs/pictograms.html.

- **Signal words:** *Danger* or *warning* will be used to emphasise hazards and indicate the relative level of severity of the hazard, assigned to a GHS hazard class and category. Some lower-level hazard categories do not use signal words. Only one signal word corresponding to the class of the most severe hazard should be used on a label.
- **Hazard statements:** Standard phrases assigned to a hazard class and category that describe the nature of the hazard. An appropriate statement for each GHS hazard should be included on the label for products possessing more than one hazard.

The additional label elements included in the GHS are:

- **Precautionary statements:** Measures to minimise or prevent adverse effects. There are four types of precautionary statements covering prevention, response in cases of accidental spillage or exposure, storage and disposal. The precautionary statements have been linked to each GHS hazard statement and type of hazard.
- **Product identifier (ingredient disclosure):** Name or number used for a hazardous product on a label or in the SDS. The GHS label for a substance should include the chemical identity of the substance. For mixtures, the label should include the chemical identities of all ingredients that contribute to acute toxicity, skin corrosion or serious eye damage, germ cell mutagenicity, carcinogenicity, reproductive toxicity, skin or respiratory sensitisation, or target organ systemic toxicity (TOST), when these hazards appear on the label.
- **Supplier identification:** The name, address and telephone number should be provided on the label.
- **Supplemental information:** Non-harmonised information on the container of a hazardous product that is not required or specified under the GHS. Supplemental information may be used to provide further detail that does not contradict or cast doubt on the validity of the standardised hazard information.

The GHS includes directions for application of the hazard communication elements on the label. In particular, it specifies for each hazard, and for each class within the hazard, what signal word, pictogram and hazard statement should be used. The GHS hazard pictograms, signal words and hazard statements should be located together on the label. The actual label format or layout is not specified in the

GHS. National authorities may choose to specify where information should appear on the label or allow supplier discretion. There has been discussion about the size of GHS pictograms and that a GHS pictogram might be confused with a transport pictogram or "diamond". Transport pictograms are different in appearance than the GHS pictograms.

The GHS has dropped the word *material* from *material safety data sheet* (MSDS); it will now be called the safety data sheet. SDSs are specifically aimed at use in the workplace. It should provide comprehensive information about the chemical product that allows employers and workers to obtain concise, relevant and accurate information that can be put in perspective with regard to the hazards, uses and risk management of the chemical product in the workplace. The SDS should contain 16 sections, as indicated in the GHS guidelines.

The primary difference between the GHS requirements in terms of headings and sections and the international industry recommendations is that Sections 2 and 3 have been reversed in order. The GHS SDS headings, sequence and content are similar to the International Organization for Standardization (ISO), EU and American National Standards Institute (ANSI) MSDS/SDS requirements.

The adoption of the GHS is expected to facilitate international trade by increasing consistency between the laws in different countries that currently have different hazard communication requirements. There is no set international implementation schedule for the GHS. The goal of the UN was broad international adoption by 2008. Different countries will require different time frames to update current regulations or implement new ones. Implementation for substances tends to be earlier than those for mixtures, which includes base oils and lubricants.

12.6.3 Types of Lubricant Package Labels

There are two main types of labels for lubricants, those for small packages (plastic bottles and tin-plate cans) and those for large packages (drums and IBCs).

For small-pack labels, the front side is the "face of the product". This will have the product's brand name, with distinctive and sophisticated product design, artwork and colouring. Examples of the front side of several different bottles of automotive engine oils are shown in Figure 12.1. Note that each of the bottles has a different shape and handling features.

With the back (rear) side of these bottles, there will be specific product information, including major international, national and original equipment manufacturers' specifications and/or approvals. The label will also include legal and disposal information. The labels will either be printed on the bottle or can using flexographic, offset or digital printing or be printed on durable adhesives.

For large-pack labels, the label will be a simple, generic design, with the lubricant's product name and grade. There may be health, safety and environmental information, although this is more likely to be included with the accompanying SDS. The drum or IBC is likely to have the marketer's distinctive colour scheme and the label will be applied to either the top or side of the drum or IBC using thermal transfer printing. Examples of distinctive large packages are shown in Figure 12.3.

12.6.4 Multilingual and Multipurpose Labels

In many regions, labels for small packs are likely now to be multilingual. That is, the information on the back of the package will be in many different languages. This enables lubricant manufacturers and marketers to supply the same product in all those countries on the label, enabling bigger print runs and lower costs. It also allows uniformity of branding. For example, Mobil 1 is known worldwide. (The front side of the pack is "neutral". often English.)

An example of a multilingual label in Europe includes English, French, German, Italian, Dutch, Danish, Swedish, Norwegian, Finnish, Polish and Czech. Obviously, with so many languages, the information given is quite brief and the printing is quite small. In Asia, an example of a multilingual label is Chinese, Japanese, Malay, Thai, Indonesian and English.

Multilingual labels are also used on occasion for commercial and industrial lubricants that are distributed widely or for slower-moving products and grades.

Marketers in some countries prefer to retain labels in their national language only, for reasons of national identity or security.

12.7 FILLING LUBRICANT PACKAGES

12.7.1 Bottle Filling

Filling plastic bottles and tin cans with lubricants is usually done only for automotive grades, notably gasoline, diesel and two-stroke engine oils, motorcycle and recreational oils, gear oils and general machine oils to be sold in the retail market.

No one type of filling machine can handle all liquids in all industries. For example, a machine that fills bottled water cannot fill cosmetic cold cream. Nor would a chemical duty filler be used to fill pharmaceutical-grade or dairy products. Although there are many different types of filling technologies, there are relatively few that are versatile, practical and cost-effective to own and operate. The choice of filling machine depends on the range of viscosities, temperature, chemical compatibility, particulate size, foam characteristics and hazardous environment considerations.

Servo pump bottle filling machines are very versatile, capable of filling nearly any type of product that can be pumped. Each nozzle has a dedicated servo-controlled pump that can deliver thin liquids, medium- and thick-viscosity liquids and liquids with large particulates. Because it is so versatile, it is often purchased by contract packagers who never know what their next filling challenge is going to be. Examples of the range of products that can be run on this machine include soaps, pharmaceutical products, oils and greases, cosmetics, salsa and sauces.

The challenges of this machine are its high capital cost and the ability of the owner to conduct normal maintenance on a more sophisticated machine. Positive displacement pumps are expensive, and so are servo-controlled drive systems. Troubleshooting and maintenance requires a reasonably competent technical-level engineer. However, if affordable, this type of filler is an outstanding choice for nearly any type of filling operation.

Net-weight bottle filling machines are best suited for liquids filled in bulk quantities (5 L) or smaller quantity products that have a very high manufactured value. Often there are products that must be sold by weight for commercial reasons, and therefore this filling machine is the only choice. Examples of this type of filler for bulk products include cleaning chemicals, enzyme solutions, oils and other medium-value products.

The operation of this type of filling machine is simple. The product bulk supply is pumped into a holding tank above a pneumatically operated valve. The valve opens and real-time net-weight information is monitored until the target weight is achieved. The valve simply shuts when the target weight is achieved. Accuracy of fills is accomplished by various "bulk and dribble" methods in the filling process so that overfills are avoided.

The advantage of this filling machine over others is that it is sometimes the only practical (and legal) type of filling for a limited range of applications and for large-volume fills. It is also very accurate and effectively provides its own quality control, assuming the weighing scale is functioning properly. The disadvantage of this type of filling machine is that it is very expensive per filling head, and it is also a relatively slow method of filling a container. For these reasons, the use of this filling technology is limited to the examples outlined above.

Automatic and semi-automatic lubricant bottle filling machines will need to be adaptable to fill different sizes and shapes of bottles. Automotive lubricants destined for retail customers are supplied in ½, 1, 2, 4 and 5 L bottles. Unless the blending plant is particularly large, with a high-volume throughput of products, it is likely to be prohibitively expensive to have a filling line dedicated to each size of bottle.

More importantly, lubricant marketers need to be able to introduce new products in differently shaped bottles, for the purposes of branding and promotion. Bottle filling lines need to be able to adjust to these changes.

Bottle filling lines, in addition to simply filling bottles, are likely to need to include bottle conveyors, capping machines, labelling machines, carton erecting machines, bottle packers and carton closing machines. At the end of these lines, the blending plant may benefit from having palletising machines and pallet wrappers.

For filling lubricant bottles, most automatic systems filling 1, 2, 4 or 5 L plastic bottles will operate at around 50 bottles per minute. Examples of different methods for filling bottles (and drums; see Section 12.7.2) are shown in Figure 12.4, and a photograph of a lubricant bottle filling line is shown in Figure 12.5. A photograph of applying labels to bottles in a lubricant bottle filling line is shown in Figure 12.6.

12.7.2 Drum and IBC Filling

Most industrial lubricants and some automotive lubricants are delivered to customers in 25 or 205 L drums and 1000 L IBCs. Engine oils delivered to car servicing garages, quick lube centres and fleet operators are often delivered in either 205 L drums or IBCs.

Drums can be filled in one of three ways:

- Bottom filling, below the surface of the oil. This is best for products which foam or vaporise during filling.

- Top filling, above the surface of the oil or grease. This is suitable for non-critical products and open-top drums and pails.
- Top surface, below the bunghole. This is the method of filling used most often for oils.

Top drum filling configurations are less expensive but limited to non-foaming bulk products. Bottom-up drum filling configurations are very flexible systems capable of filling almost any bulk product within a wide range of foamy characteristics and viscosities. Systems can be designed to fill drums singly, in sequence, or pallets

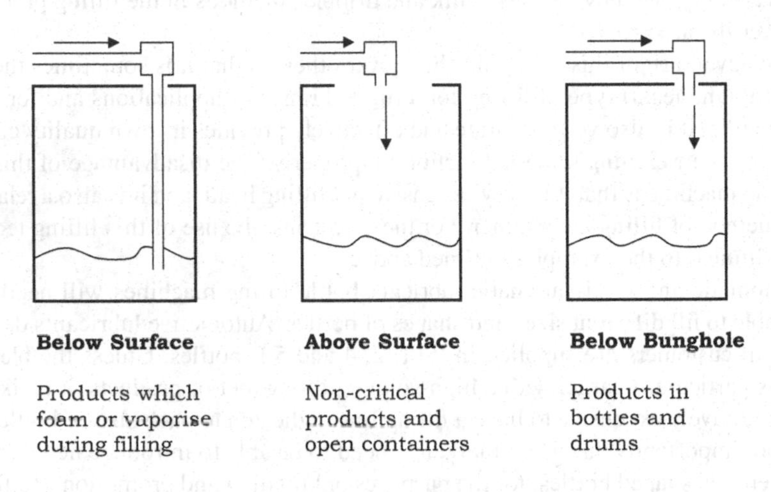

FIGURE 12.4 Lubricant bottle and drum filling methods.

FIGURE 12.5 Lubricant bottle filling line. (From Fuchs Lubricants [UK] plc. With permission.)

FIGURE 12.6 Applying labels in a lubricant bottle filling line. (From Fuchs Lubricants [UK] plc. With permission.)

of four drums, with either one, two or four heads. Some systems are able to accept either configuration.

Depending on the product and process, either an electronic weigh scale or a mass flow meter is used for measuring quantities of oil to be filled. Filling systems can be automatic or semi-automatic.

With automatic systems, a video recognition system locates the positioning of the bunghole, the dosing arm and nozzle are inserted into the bunghole and pumping is initiated to dispense the oil. They have an articulated balancing assist arm, with controls to move and park the fill head, including mechanisms for nozzle height, nozzle park/unpark, nozzle dive, nozzle open and pump/supply on/off. The weighing scale is usually stainless steel, with a remotely located digital readout and integration of the scale's operation with filler controls. Capping and crimping can also be automated and the filling zone can be protected. This solution gives the operator more time for other tasks.

With a semi-automatic system, the operator positions the dosing nozzle just above the bunghole of the drum to be filled. The machine has an automatic anti-drop system. These systems tend to be slightly more flexible, accommodating drums and IBCs from 25 to 1000 L capacities. They may also have a pallet conveyor to optimise the station, often an operator-actuated powered roller conveyor over or on scale for drum pallet control (load/unload).

Both top-fill and bottom-fill drum filling systems use an integrated weighing scale and powered pallet roller conveyor. The system operates with minimal operator involvement. The operator positions a pallet or drums on the scale platform with a conveyor control switch. They then use the nozzle balancing arm switch for manual (zero-weight) positioning to the drum or IBC opening. The operator then activates autofill and the nozzle opens and the pump fills the drum or IBC to the correct weight.

Most modern systems have a programmable logic controller (PLC), coarse and fine options, data logging and pneumatic positioners and can be operated in either automatic, semi-automatic or manual modes. Some have single nozzles, while some have multiple heads. Some drum filling systems are supplied with custom pump systems or can be fed by the blending plant's existing bulk supply system.

The degree of automation of a drum filling station can be selected by the blending plant's management. Any or all of the following sequence of operations can now be automated: drum storage, de-nesting, infeed, positioning, filling, labelling, printing, membrane sealing, closing, palletising and outfeed. A fully automatic drum filling line will be able to fill up to 125×205 L drums per hour.

All the equipment on a drum filling system should be robust to the occasional bump from a forklift truck, but most importantly, the conveyors at the start and end of the filling line. A photograph of an individual drum filling unit is shown in Figure 12.7, and a photograph of several drum filling lines in a lubricant blending plant is shown in Figure 12.8. An automatic drum filling machine is shown in Figure 12.9.

12.7.3 Road Tanker, Rail Tank Wagon and ISOTainer Filling

Lubricants can be delivered to customers in bulk quantities in road tankers, rail tank wagons, ISOTainers (also called ISOTanks or ISO tank containers) or flexi-bags

FIGURE 12.7 Lubricant drum filling unit. (From Total Lubmarine. With permission.)

FIGURE 12.8 Multiple drum filling lines in a lubricant blending plant. (From Total Lubmarine. With permission.)

FIGURE 12.9 Automatic drum filling machine. (From Fuchs Lubricants [UK] plc. With permission.)

inside standard 20-foot shipping containers. (Base oils and additives can also be delivered to lubricant blending plants in these containers.)

A road tanker is a motor vehicle designed to carry liquefied loads, dry bulk cargo or gases on roads. Many variants exist due to the wide variety of liquids that can be transported. The tanks on road tankers used for lubricants, base oils or additives are generally made of stainless steel and are likely to have three, four or five compartments, each with their own filling hatch and outlet delivery pipe. Some road tankers are semi-trailer trucks.

Road tankers are described by their size or volume capacity. Large tankers typically have capacities ranging from 20,000 to 44,000 L. They are distinguished by shape, usually a cylindrical tank lying horizontally on the chassis of the tanker.

Rail tank wagons are essentially the same as road tankers, except that the tank is mounted on the chassis of a railway wagon. Their capacities range from 40 to 100 tonnes, with the smaller types having two axles and the larger types four axles in a double-bogie arrangement. In the oil industry, the smaller rail tank wagons tend to be used to transport base oils, lubricants and petrochemicals, while the bigger rail tank wagons tend to be used to transport crude oil, gasoline, diesel or aviation fuels.

ISOTainers (ISOTanks) are tanks that are constructed to the ISO standards for shipping containers, making them suitable for different modes of transportation (ships, trucks and railway wagons). Both hazardous and non-hazardous products can be transported in ISOTainers. A photograph of an ISOTainer is shown in Figure 12.10.

An ISOTainer is a vessel of stainless steel surrounded by an insulation and protective layer of usually polyurethane and aluminium. The vessel is in the middle of a steel frame. The frame is made according to ISO standards and is 19.8556 feet (6.05 m) long, 7.874 feet (2.40 m) wide and 7.874 feet (2.40 m) or 8.374 feet (2.55 m) high. The contents of the tank range from 17,500 to 26,000 L.

Flexi-bags (also called flexi-tanks) offer an alternative to ISOTainers for shipping all types of non-hazardous liquids, including base oils and lubricants. Flexi-bags are usually loaded into standard 20-foot food-grade ISO containers for sea, road and/or rail transport. Sizes range from 16,000 to 26,000 L.

The bags are constructed from multiple layers of food-grade polyethylene and enclosed in an outer layer of high-strength abrasion-resistant woven PP. They are designed for single-trip use, are environmentally friendly and can be recycled.

FIGURE 12.10 Photograph of an ISOTainer. (From M&S Logistics Ltd. With permission.)

Different types of valves can be fitted to a flexi-bag, to suit customers' requirements. Top or bottom loading and unloading options are also available.

The empty flexi-bag is unrolled inside the open container and filled with liquid product, and the container is closed. On arrival at the container's destination, the product is pumped out of the flexi-bag, which can then be rolled up and returned for reuse as required. The by now empty container can then be used for a return journey to transport other cargo. Some manufacturers of flexi-bags customise them for specific product types, so that the product can be handled by the customer's equipment and staff.

The claimed advantages of flexi-bags are no cleaning of empty tankers or ISOTainers, lower total freight costs (resulting from the use of the container on the return journey, as tankers and ISOTainers may have to be returned empty) and the inner liner is reuseable, recyclable or disposable.

Depending on the product, many ISO tank container operators argue that flexi-bags can be more costly and less environmentally sound. This is generally the case for higher-value products and/or environmentally dangerous products, because the transport protection offered by a flexi-bag is less and the level of product that is left in the flexi-bag (2% to 5%) is higher than the amount left in an ISOTainer.

Filling road tankers, rail tank wagons, ISOTainers and flexi-bags with finished lubricants to be delivered in bulk involves the same principles and methods as transferring products to and from storage tanks.

Pipelines should be piggable and couplings to tankers and wagons should be secure and sealed from ingress of moisture and dirt. It is also advisable to locate the filling facilities under a fixed roof in a dedicated loading gantry, particularly in wet, hot or very dry climates. A photograph of a suitable road tanker loading gantry is shown in Figure 12.11.

FIGURE 12.11 Photograph of a road tanker filling gantry.

12.8 SUMMARY

Packaging and filling lubricants in a blending plant is just as important as the blending processes themselves.

Improperly filled, poorly sealed, incorrectly labelled or damaged bottles, drums or IBCs are almost certain to lead to customer dissatisfaction and complaints. Correctly filled packages are intended to protect both the contents and the brand image of the lubricant manufacturer.

13 Lubricant Storage

13.1 INTRODUCTION

Lubricants can be stored in tanks, drums, small containers or cans. The requirements in each case are that the material should be protected from contamination and should not deteriorate in storage, and account has been taken of health, safety and environmental matters. For drums and small containers, it is important that the contents are easily identifiable from the markings and that all grades are easily accessible for shipping purposes.

Where these tanks, drums and small containers are stored is also very important for the protection of the lubricants. Although the design, facilities and operation of a blending plant warehouse share many of the features of warehouses in other industries, a few specific requirements are necessary.

13.2 STORAGE VESSELS AND CONTAINERS

13.2.1 Bulk Storage in Tanks

Fixed-roof tanks of various sizes may be used to store lubricants. Before use they should be thoroughly cleaned, should be scale-free and may be coated internally with a proprietary protective coating that is of a type compatible with the lubricant to be stored. The tank should initially be dry, and filling arrangements must be such that ingress of water to the product is eliminated.

Because base oils may contain a little water, or condensation may take place in moist climates with fluctuating temperatures, and in case of accidents, the bottom of a base oil storage tank should be sloping and provided with a drain from which water and any bottom residues can be removed.

For viscous base oils and finished lubricants in cooler climates, it will be necessary to provide some heating. For lighter oils, an outflow (suction) heater is sufficient, but if bulk heating is necessary the heating coils must be sound and should be fed preferably with hot oil or water, or with only low-pressure or exhaust steam to prevent local overheating.

The tanks should stand on an impervious concrete platform and be bunded (provided with a catchment area for the product, should the tank leak). Watch must be kept for leaks at valves and flanges. Leaks should be eradicated as soon as possible. Lagging (insulation) of transfer lines is common, but the tanks themselves are not normally lagged for finished lubricants. However, regular use of viscous base oils may justify lagging to save energy.

If horizontal-type tanks on piers are used for base oils, the same general remarks apply, with the tanks being slightly sloping toward a draw-off point for water and bottom samples. Where draw-off pipelines are common to different grades, a pigging system should be installed. With horizontal tanks for storing finished lubricants, the

FIGURE 13.1 Photograph of tanks in a lubricant blending plant. (From Total Lubmarine. With permission.)

tank should also be slightly sloping toward a draw-off point, so that the tank can be drained completely, if necessary.

A photograph of the range of tanks used to store base oils, bulk additives and blended lubricants in a typical lubricant blending plant is shown in Figure 13.1.

13.2.2 Drum Storage

Larger volumes of finished lubricants are often stored in drums. This is the most difficult and potentially the most hazardous form of storage. Ideally, drums should be stored on their sides and with the bungs below the liquid level. This is to prevent water from collecting in the tops of the rims and being drawn in as the drums cool. Keeping the bung seals moistened with product guards against leakage. However, storing large numbers of drums horizontally is hazardous unless a large amount of special-purpose racking is used, which can be costly. Drums stored one on top of the other should never be more than three drums high, and the ends must be securely chocked. Products must be stacked so that access to the lower layer is not required until the first layer has been removed. In no cases should drums be directly laid on the ground, but they should be on battens or an impervious base. Provision should be made for stock rotation on, or as close to, a first-in-first-out (FIFO) basis as can be achieved.

The safe storage period or "shelf life" of products must be considered, and product should not be left for excessive periods at the bottom of a stack.

Drums are frequently stored in fours on pallets. Full drums should normally be stored no more than two pallets high, although empty drums can be stored in higher piles if adequate care is taken in placing and removing the top layers. In particular, forklift trucks should have strong safety cabins and not be of the open type. Vertically stored drums must be protected from rain, and if they cannot be stored in a warehouse, they should be sheeted over with tarpaulins or plastic wrapped. In hot climates, the drums should be protected from direct sunlight by light-coloured screens or roofing.

Leakage must be prevented as much as possible. If taps are used on drums, then drip trays must be provided and care taken that the taps are functioning properly and are shut off after use.

Small hand pumps are sometimes used to withdraw material from a drum placed on end, but again care must be taken that this does not leave a trail of oil when removed from the drum. To mop up accidental oil spills, proprietary crystalline materials are available which are much less hazardous than traditional sawdust.

Care must be taken when removing drums with forklift trucks that the drums are not accidentally punctured, thereby producing both an environmental problem and a health risk. Drums are often returnable or reusable and therefore care should be taken at all times not to damage them unduly in handling operations.

13.2.3 Storage in Intermediate Bulk Containers

In many countries, automotive engine oils are now delivered to retail lubricant outlets, such as car and truck dealerships, fleet operators and maintenance garages, in intermediate bulk containers (IBCs). Some industrial lubricants, particularly hydraulic oils, gear oils and compressor oils, are also delivered to larger customers in IBCs.

Storage of lubricants in IBCs is similar to storage in plastic bottles (see Section 13.2.4), in that ingress of moisture is much less likely than it is with steel or plastic drums. However, bungs on top of IBCs should be kept tight, the tops of IBCs should ideally be covered and it is preferable to store IBCs inside.

13.2.4 Plastic Bottles and Tin-Plate Cans

Empty plastic bottles and tin-plate cans should always be kept under cover in a warehouse or other suitable building. Small containers will normally be packed in multiples in cardboard containers. Where the stock is for in-house use, provision must be made for unpacking and disposing of cartons, and not allowing these to become oil soaked where they will present a hazard.

Because tin cans and plastic bottles are lightweight and much less robust than steel drums, they cannot be stacked on top of each other either very easily or very high. Even cardboard cartons of identically sized plastic bottles cannot be stacked more than three or four high. This means that pallets of plastic bottle storage cartons cannot be stacked on top of each other. This places limitations on the design and operation of lubricant storage warehouses.

In modern lubricant blending plants, it has become common to have a plastic bottle blow moulding facility in the lubricant filling area. Transporting lots of empty plastic bottles to a blending plant uses a great amount of energy when compared with making the bottles on site, as the trucks delivering the empty bottles are transporting a lot of fresh air.

13.3 LUBRICANT STORAGE

13.3.1 Siting the Lubricants Store

The ideal site for a lubricants store in a blending plant has:

- A good reception area, with free access for vehicles and ample room for loading.
- Adequate space for filled packages (drums, cans and plastic bottles) of all grades and sizes of packed products.
- Adequate space for empty packages (drums and polycontainers) near the unloading point, so they can be delivered by delivery vehicles.
- A well-equipped loading dock, with direct access to the oil store.
- A location that minimises the work needed to get lubricants into and out of the store.

In a blending plant, the storage warehouse for finished, packed lubricants is most usually either part of the main building that houses the blending and filling equipment or immediately adjacent to it.

13.3.2 Indoor Storage

Indoor storage of lubricants is desirable at all times. Since most companies cannot store everything indoors, however, they have to make the best use of the space available. They require easy access to the stock and freedom to use the packages in the order delivered. Three methods of storing packages are available and in common use: free stacking, palletisation and the use of racks.

In free stacking, the packages are placed on top of one another, and the safe height of the stack depends on its stability and the weight that the lower packages can support. The smallest packages, cans of 1 gallon (4 to 5 L) or less, and plastic bottles are usually packed in stout fibreboard cases, or they can be bound together with tape for stability. Five-gallon (23-litre) drums and grease kegs are usually handled singly (unless palletised) and can be manhandled. The movement of drums calls for at least a hand trolley and skid, and preferably a forklift device or mechanical hoist. The use of planks or slatted frames stabilises the stack and helps to prevent damage to the lower layers.

If a forklift truck or similar is available, palletisation eases stacking, reduces handling risks and allows better access to the lower layers.

Small drums can be stacked on pallets in tiers seven high, though such a practice is unusual except in a central store serving a number of secondary sites, or where space limitations justify frequent "breaking" of the stacks. Drums also can be palletised, but more often they are stacked with interposed strips of timber or heavy-duty recycled plastic.

Steel racks, of slotted or plain angle, or of the clamped tubular type, have several merits; they allow space to be used to the best advantage, they ease stock handling and they encourage regular turnover. They should be installed with aisles wide enough to allow a forklift truck to be manoeuvred.

Only in very cold climates can indoor temperatures drop low enough to produce adverse effects in a lubricant. At the other end of the temperature scale, though, excessive heat due to the proximity of steam pipes, boilers, furnaces or flues should be avoided for grades containing volatile solvents. In many cases, because of insurance requirements or local fire regulations, it may be necessary to house such products, together with kerosene, white spirit and so on, in a store separate from lubricants. If one part of the store is hot, it should be reserved for oils of high viscosity.

The store should be dry at all times. Those containers which are made of painted sheet steel or tin plate, will corrode (rust) if left in a damp condition for long enough. Tin plate in particular can rust through in only a few weeks, and the result is that both water and abrasive rust enter the lubricant. In severe cases, rust can obscure the grade-name markings on tin-plate cans.

Well-designed racks save space, ease handling and ensure a FIFO procedure. In the absence of racks, palletising aids handling, but bottom layers tend to remain undisturbed.

If indoor storage is limited, it should be reserved for small packages and for lubricants affected by frost and heat. Special precautions should be taken with outdoor stock; it is desirable to arrange for a short stock life or frequent turnover.

Do not use empty steel drums for road barriers or scaffold-pole support. In particular, *never* use them as work tables or trestles for welding or brazing work, because of the risk of explosions.

13.3.3 Outdoor Storage

Drums should not be stored upright outdoors unless they are upside down, with the bungs at the bottom. Gaskets and O-ring seals around bungs are rarely airtight. Any rainwater that collects on the tops of drums stored upright is usually sucked into the drum as it "breathes" (expansion and contraction of the drum's contents) due to daily changes in air temperatures.

Gaskets and O-ring seals around bungs in drums should ideally be wetted by the drum's contents. Drums stored horizontally should be placed so that their bungs are below the level of the oil, ideally at the three o'clock and nine o'clock positions.

Drums stored on their sides should be clear of the ground, perhaps on baulks (square logs) of timber (wood) or heavy-duty recycled plastic. They can be stacked three high in this way but must be carefully wedged to prevent movement. Steel sections are sometimes used instead of timber or plastic baulks.

All too frequently, when drums are stacked, the top ones are used and quickly replaced by new ones, so the lower ones remain undisturbed for months or even years. For this reason alone, racking is to be preferred. Sloping racks, in which drums are loaded at one side and removed from the other (FIFO), can be used. In addition to ensuring regular replacement, a rack is convenient to load, is safe and makes for reasonable use of space. A corrugated-iron or plastic roof over the rack also provides a degree of protection against rain. Because of the rolling of drums, however, the ideal positions of the bungs cannot be maintained for long.

If small packages, drums and grease kegs have to be stored outdoors, they should be covered over with plastic or waterproof sheeting, with free access for air, and should be examined regularly. The size of the stock should be adjusted to provide a rapid turnover. The containers should not stand directly on the ground. Rather, they should be raised so that air can circulate beneath and around them.

If possible, outdoor storage sites should not be near dusty areas such as quarries and unmade roads, because thick dust on containers is likely to contaminate the contents when they are opened.

13.3.4 Ingress of Moisture

All packages need protection from rainfall and condensation, partly because of the dangers of contamination. Although the paint on containers is of high quality, it can flake off after prolonged exposure, particularly to corrosive atmospheres. This leads to obliteration of the markings and eventually to rusting of the outer surfaces, which in turn could progress far enough to cause serious contamination and loss of the contents.

Most new lubricants have a moisture content of less than 50 parts per million (0.005%); transformer and refrigerator oils are required to have even less. Care must be taken to prevent water entering all packages, and special precautions are needed to guard against the ingress of moisture as the result of expansion and contraction with daily heat and cold.

13.3.5 Storage of Special Types of Lubricant

Tanks for electrical and refrigerator oils are usually lined with amine-cured epoxy resin, and the air vents are protected with a silica-gel breather, to remove moisture. Self-sealing couplings are fitted to fill pipes. During manufacture, some tanks for white oils are internally shot-blasted and immediately coated with rust-preventive oil. After erection on site, they are swabbed internally with the oil to be stored. Desiccators are seldom needed in the storage of these oils, but as a precaution the air vents on the tanks are protected with filters.

13.4 BLENDING PLANT WAREHOUSE

A warehouse is a commercial building for storage of goods. Warehouses are used by manufacturers, importers, exporters, wholesalers, transport businesses and other organisations. They are usually large plain buildings, without windows but with excellent ventilation. Stored goods can include any raw materials, components or finished products.

The main processes in a warehousing include:

- Receiving.
- Putting away.

- Order preparation/picking.
- Shipping.
- Inventory management (cycle counting, addressing and more).

Some of the most common warehouse storage systems are:

- Pallet racks, including selective, drive-in, drive-through, double-deep, pushback and gravity flow types.
- Mezzanine flooring, including structural, roll formed, rack supported and shelf supported.
- Industrial shelving, including metal, steel, wire and catwalk types.
- Automated storage and retrieval system (ASRS), including vertical carousels, vertical lift modules (VLMs), horizontal carousels and robotic stackers.

The finished lubricants warehouse in a blending plant will often be the largest building or occupy the largest space. The production capacity of a blending plant and the size of the warehouse are inextricably linked, since the warehouse will usually need to accommodate between 6 and 8 weeks of stock of lower offtake products. A photograph of a typical modern lubricant blending plant warehouse is shown in Figure 13.2.

FIGURE 13.2 Lubricant blending plant warehouse. (From Fuchs Lubricants [UK] plc. With permission.)

Traditional warehousing has declined since the last decades of the twentieth century, with the gradual introduction of just-in-time (JIT) methodologies. The JIT system promotes product delivery directly from suppliers to consumers with minimal or no use of warehouses. However, with the gradual implementation of offshore outsourcing and offshoring in about the same time period, the distances between manufacturers and their customers have increased considerably in many countries and regions, necessitating at least one warehouse per country or per region in any typical supply chain for a given range of products.

13.5 WAREHOUSE MANAGEMENT AND AUTOMATION

13.5.1 Warehouse Management Systems

Material direction and tracking in a warehouse can be coordinated by a warehouse management system (WMS), a database-driven computer programme. Logistics personnel use the WMS to improve warehouse efficiency by directing putaways and to maintain accurate inventory by recording warehouse transactions.

A WMS is a key part of the supply chain and primarily aims to control the movement and storage of materials within a warehouse and process the associated transactions, including shipping, receiving, putaway and picking. The systems also direct and optimise stock putaway based on real-time information about the status of bin or rack utilisation.

WMSs often use auto ID data capture (AIDC) technology, such as barcode scanners, mobile computers, wireless local area networks (LANs) and potentially radio frequency identification (RFID) to efficiently monitor the flow of products. Once data have been collected, there is either a batch synchronisation with or a real-time wireless transmission to a central database. The database can then provide useful reports about the status of goods in the warehouse.

The objective of a WMS is to:

- Provide a set of computerised procedures to handle the receipt of stock and returns into a warehouse facility.
- Model and manage the logical representation of the physical storage facilities, for example, racking.
- Manage the stock within the facility.
- Enable a seamless link to order processing and logistics management in order to pick, pack and ship products out of the facility.

WMSs can be stand-alone systems or modules of an enterprise resource planning (ERP) system or supply chain execution suite.

The primary purpose of a WMS is to control the movement and storage of materials within a warehouse. It might even be described as the legs at the end of the line which automate the store, traffic and shipping management.

In its simplest form, the WMS can data track products during the production process and act as an interpreter and message buffer between existing ERP systems and WMSs. Warehouse management is not just managing within the boundaries of a warehouse today; it is much wider and goes beyond the physical boundaries. Inventory management, inventory planning, cost management, information technology (IT) applications and communication technology to be used are all related to warehouse management. The container storage, loading and unloading are also covered by warehouse management today.

Warehouse management is now part of supply chain management (SCM) and demand management. Even production management is to a great extent dependent on warehouse management. Efficient warehouse management gives a cutting edge to any organisation's supply chain logistics. Warehouse management does not just start with receipt of material but actually starts with initial planning when the container design is made for a product. Warehouse design and process design within the warehouse (for example, wave picking) are also part of warehouse management.

Warehouse management monitors the progress of products through the warehouse. It involves the physical warehouse infrastructure, tracking systems and communication between product stations.

An automated storage and retrieval system (ASRS) consists of a variety of computer-controlled methods for automatically placing and retrieving loads from specific storage locations. ASRSs are typically used in applications where:

- There is a very high volume of loads being moved into and out of storage.
- Storage density is important because of space constraints.
- No value-adding content is present in this process.
- Accuracy is critical because of potential expensive damage to the products.

ASRSs are categorised into three main types: single masted, double masted and man-aboard. Most are supported on a track and ceiling guided at the top by guide rails or channels to ensure accurate vertical alignment, although some are suspended from the ceiling. The "shuttles" that make up the system travel between fixed storage shelves to deposit or retrieve a requested load (ranging from a single book in a library system to a several-tonne pallet of goods in a warehouse system). As well as moving along the ground, the shuttles are able to telescope up to the necessary height to reach the load, and can store or retrieve loads that are several positions deep in the shelving. A photograph of a typical ASRS in a lubricant blending plant is shown in Figure 13.3.

To provide a method for accomplishing throughput to and from the ASRS and the supporting transportation system, stations are provided to precisely position inbound and outbound loads for pickup and delivery by the crane.

A man-aboard ASRS offers significant floor space savings. This is because the storage system heights are no longer limited by the reach height of the order picker. Shelves or storage cabinets can be stacked as high as floor loading, weight capacity, throughput requirements and/or ceiling heights will permit. Man-aboard ASRSs are far and away the most expensive picker-to-stock equipment alternative. Aisle-captive storage/retrieval machines reaching heights up to 40 feet cost around $125,000. Hence, there must be enough storage density and/or productivity improvement over cart and tote picking to justify the investment. Also, because vertical travel is slow compared with horizontal travel, typical picking rates in man-aboard operations

FIGURE 13.3 Lubricant blending plant automated storage and retrieval system. (From Total Lubmarine. With permission.)

range between 40 and 250 products per person-hour. The range is large because there are a wide variety of operating schemes for man-aboard systems. Man-aboard systems are typically appropriate for slow-moving items where space is fairly expensive.

The VLM is a computer-controlled automated vertical lift, storage and retrieval system. Functionally, stock within the VLM remains stationary on front and rear tray locations. On request, a movable extractor unit travels vertically between the two columns of trays and pulls the requested pallet from its location and brings it to an access point. The operator then picks or replenishes stock, and the tray is returned to its home. The VLM system offers variable tray sizes and loads, which could be applied in different industries, logistics and office settings. The VLM systems could be customised to fully use the height of the facility, even through multiple floors. With the capability of multiple access openings on different floors, the VLS system is able to provide an innovative storage and retrieval solution. The rapid movement of the extractor as well as the integrated warehouse management software system can dramatically increase the efficiency of the picking process. Unlike large ASRSs, which require a complete overhaul of the warehouse or production line, the VLMs are modularised, and so can be easily integrated into an existing system, or rolled out gradually over different phases.

The evolution of WMSs is very similar to that of many other software solutions. Initially a system to control movement and storage of materials within a warehouse, the role of WMS is expanding to include light manufacturing, transportation management, order management and complete accounting systems. To use the grandfather of operations-related software as a comparison, material requirements planning (MRP) started as a system for planning raw material requirements in a manufacturing environment. Soon MRP evolved into manufacturing resource planning (MRPII), which took the basic MRP system and added scheduling and capacity planning logic. Eventually, MRPII evolved into ERP, incorporating all the MRPII functionality with full financials and customer and vendor management functionality.

Whether WMS evolving into a warehouse-focused ERP system is a good thing is up to debate. What is clear is that the expansion of the overlap in functionality between WMS, ERP, distribution requirements planning, transportation management systems, supply chain planning, advanced planning and scheduling and manufacturing execution systems will only increase the level of confusion among companies looking for software solutions for their operations.

13.5.2 Advantages and Disadvantages of WMS

Even though WMS continues to gain added functionality, the initial core functionality of a WMS has not really changed. The primary purpose of a WMS is to control the movement and storage of materials within an operation and process the associated transactions. Directed picking, directed replenishment and directed putaway are the key to WMS. The detailed set-up and processing within a WMS can vary significantly from one software vendor to another. However, the basic logic will use a combination of item, location, quantity, unit of measure and order information to determine where to stock, where to pick and in what sequence to perform these operations.

Not every warehouse needs a WMS. Certainly, any warehouse could benefit from some of the functionality, but the benefits may not be big enough to justify the initial and ongoing costs associated with WMS. WMSs are big, complex, data-intensive applications. They tend to require a great deal of initial set-up, significant system resources to run and considerable ongoing data management to continue to run. An organisation needs to “manage” its WMS. Large organisations will frequently end up creating a new information systems (IS) department with the sole responsibility of managing the WMS.

Vendors of WMS sometimes claim that the system(s) will:

- Reduce inventory.
- Reduce labour costs.
- Increase storage capacity.
- Improve customer service.
- Increase inventory accuracy.

In reality, the implementation of a WMS (together with automated data collection [ADC]) is likely to give an organisation an increase in accuracy, reductions in labour costs (provided that the labour required to maintain the system is less than the labour saved on the warehouse floor), and a greater ability to service the customer by reducing cycle times. Expectations of inventory reduction and increased storage capacity are less likely. While increased accuracy and efficiencies in the receiving process may reduce the level of safety stock required, the impact of this reduction will likely be negligible in comparison with overall inventory levels. The predominant factors that control inventory levels are lot sizing, lead times and demand variability. It is unlikely that a WMS will have a significant impact on any of these factors. And while a WMS certainly provides the tools for more organised storage which may result in increased storage capacity, this improvement will be relative to just how sloppy the organisations were pre-WMS processes.

Beyond labour efficiencies, the determining factors in deciding to implement a WMS tend to be more often associated with the need to do something to service the customers that the current system does not support (or does not support well). These include FIFO, cross-docking, automated pick replenishment, wave picking, lot tracking, yard management, ADC and automated material handling equipment.

The set-up requirements of WMS can be extensive. The characteristics of each item and location must be maintained either at the detail level or by grouping similar items and locations into categories. Example item characteristics at the detail level include:

- Exact dimensions and weight of each item in each unit of measure the item is stocked. In the case of a lubricants storage warehouse, this will include bottles, cartons, cans, drums and pallets.
- Information about whether the containers can be mixed, such as cartons of different products in plastic bottles.
- Whether the products are rackable.
- The maximum stack height.
- The maximum quantity per location.
- Hazard classifications.
- Finished goods or raw materials.
- Fast-moving versus slow-moving products.

Although some operations will need to set up each item this way, most operations will benefit by creating groups of similar products. For example, in a lubricants storage warehouse, cartons of 1 L plastic bottles and cartons of 5 L plastic bottles would be grouped separately. This would also apply to 25 and 205 L drums. WMS planners might also create groups for the different types of locations within the warehouse. The carton storage area might be separated from the drum storage area.

13.5.3 WMS Operation

If the operation of a WMS sounds simple, the reality is that most operations have a much more diverse product mix and will require much more system set-up. And setting up the physical characteristics of the product and locations is only part of the picture. Having set up enough so that the system knows where a product can fit and how many will fit in that location, the planners now need to set up the information needed to let the system decide exactly which location to pick from, replenish from or to, and putaway to. The sequence in which events should occur (remember, WMS is all about "directed" movement) must also be determined. All this is done by assigning specific logic to the various combinations of item/order/quantity/location information that will occur.

Location sequence is the simplest logic. A flow of products through the warehouse is defined and a sequence number is assigned to each location. In order picking, this is used to sequence the picks to flow through the warehouse. In putaway, the logic would look for the first location in the sequence in which the product would fit.

The next step is zone logic. By breaking down the storage locations into zones, the WMS can direct picking, putaway or replenishment to or from specific areas of the warehouse. Since zone logic only designates an area, it will need to be combined with some other type of logic to determine exact location within the zone.

Fixed location logic uses predetermined fixed locations per item in picking, putaway and replenishment. Fixed locations are most often used as the primary picking location in piece-pick and case-pick operations. However, they can also be used for secondary storage.

Random location logic is a slightly misleading description, since computers cannot be truly random (nor would an organisation want them to be). Random locations generally refer to areas where products are not stored in designated fixed locations. Like zone logic, WMS planners will need some additional logic to determine exact locations.

FIFO directs picking from the oldest inventory first. The opposite of FIFO, last-in-first-out (LIFO) only has applications for perishable goods that are sold and transported internationally. Organisations that market perishable goods can use LIFO for overseas customers (because of longer in-transit times) and FIFO for domestic customers.

Quantity or unit of measure allows the WMS to direct picking from different locations of the same item based on the quantity or unit of measure ordered. For example, pick quantities less than 25 units would pick directly from the primary picking location while quantities greater than 25 would pick from reserve storage locations.

Fewest locations logic is used primarily for productivity. Pick-from-fewest logic will use quantity information to determine the least number of locations needed to pick the entire pick quantity. Put-to-fewest logic will attempt to direct putaway to the fewest number of locations needed to stock the entire quantity. While this logic appears impressive from a productivity standpoint, it generally results in very poor space utilisation. The pick-from-fewest logic will leave small quantities of an item scattered all over the warehouse, and the put-to-fewest logic will ignore small and partially used locations.

Pick-to-clear logic directs picking to the locations with the smallest quantities on hand. This logic is excellent for space utilisation.

Reserved locations logic is used when the organisation wants to predetermine specific locations to putaway to or pick from. An application for reserved locations would be cross-docking, where the organisation may specify that certain quantities of an inbound shipment be moved to specific outbound staging locations or directly to an awaiting outbound trailer.

Nearest location is also called proximity picking/putaway. This logic looks to the closest available location of the previous putaway or pick. WMS planners need to look at the set-up and test this type of logic to verify that it is picking the shortest route and not the actual nearest location. The shortest distance between two points is a straight line. The logic may pick a location 30 feet away (believing it is closest) that requires the worker to travel 200 feet up and down aisles to get to it while there was another available location 50 feet away in the same aisle.

Maximise cube logic is found in most WMSs. However it is seldom used. Cube logic basically uses unit dimensions to calculate the cube (cubic inches per unit) and then compares them with the cube capacity of the location to determine how much will fit. If the units are capable of being stacked into the location in a manner that fills every cubic inch of space in the location, cube logic will work. Since this rarely happens in the real world, cube logic tends to be impractical.

Consolidate logic looks to see if there is already a location with the same product stored in it with available capacity. This may also create additional moves to consolidate like product stored in multiple locations.

Batch sequence logic is used for picking or replenishment. It will use the batch number or date to determine locations to pick from or replenish from.

It is very common to combine multiple logic methods to determine the best location. For example, an organisation may choose to use pick-to-clear logic within FIFO logic when there are multiple locations with the same receipt date. It may also change the logic based on current workload. During busy periods, the organisation may chose logic that optimises productivity, while during slower periods it could switch to logic that optimises space utilisation.

13.5.4 Other WMS Functionality and Considerations

Support for various picking methods (wave picking, batch picking and zone picking) varies from one system to another. In high-volume fulfilment operations, picking logic can be a critical factor in WMS selection.

Task interleaving describes functionality that mixes dissimilar tasks such as picking and putaway to obtain maximum productivity. Used primarily in full-pallet-load operations, task interleaving will direct a lift truck operator to put away a pallet on his or her way to the next pick. In large warehouses, this can greatly reduce travel time, not only increasing productivity but also reducing wear on the lift trucks and saving on energy costs by reducing lift truck energy use. Task interleaving is also used with cycle-counting programmes to coordinate a cycle count with a picking or putaway task.

It is generally assumed that when an organisation implements a WMS, it will also be implementing ADC, usually in the form of radio frequency (RF) portable terminals with barcode scanners. Many WMS advisors recommend incorporating the ADC hardware selection and the software selection into a single process. This is especially true if the organisation is planning on incorporating alternate technologies, such as voice systems, RFID or light-directed systems. It may be that a higher-priced WMS package will actually be less expensive in the end since it has a greater level of support for the types of ADC hardware that will be used.

In researching WMS packages, an organisation may see references like "supports", "easily integrates with", "works with", and "seamlessly interfaces with" in describing the software's functionality related to ADC. Since these statements can mean just about anything, it is important to ask specific questions related to exactly how the WMS has been programmed to accommodate ADC equipment. Some WMS products have created specific versions of programmes designed to interface with specific ADC devices from specific manufacturers. If this WMS/ADC device combination works for the warehouse operation, some programming or set-up time can be saved. If the WMS does not have this specific functionality, it does not mean that the organisation should not buy the system; it just means that it will have to do some programming either on the WMS or on the ADC devices. Since programming costs can easily put the system over budget, an estimate of these costs should be investigated in the initial planning phases. As long as the organisation is working closely with the WMS vendor and the ADC hardware supplier at an early stage in the process, it should be possible to avoid any major surprises.

If an organisation is planning to use automated material handling equipment, such as carousels, ASRS units, AGVs (automated guided vehicles), pick-to-light systems or sortation systems, it may want to consider integrating these during the software selection process. Since these types of automation are very expensive and are usually a core component of a warehouse, an organisation may find that the equipment will drive the selection of the WMS. As with ADC, planners should be working closely with the equipment manufacturers during the software selection process.

If an organisation's suppliers are able to send advanced shipment notifications (ASNs), preferably electronically, and attach compliance labels to the shipments, it may be wise to make sure that the WMS can use this to automate the receiving process. In addition, if the organisation has requirements to provide ASNs for customers, it will also want to verify this functionality.

Most WMSs will have some cycle-counting functionality. Modifications to cycle-counting systems are common to meet specific operational needs.

Cross-docking is the action of unloading materials from an incoming trailer or rail car and immediately loading these materials in outbound trailers or rail cars, thus eliminating the need for warehousing (storage). In reality, pure cross-docking is less common. Most cross-docking operations require large staging areas where inbound materials are sorted, consolidated and stored until the outbound shipment is complete and ready to ship. If cross-docking is part of a warehouse operation, verifying the logic the WMS uses to facilitate this will be required.

In practice, most lubricant blending plant warehouses do not use cross-docking, unless some finished products, such as synthetic refrigerator compressor oils, are brought in from sub-suppliers for onward shipment to the lubricant marketing company's supply chain. In most cases, purchased-in lubricants will be sent directly to distribution warehouses.

For parcel shippers, pick-to-carton logic uses item dimensions and weights to select the shipping carton prior to the order-picking process. Items are then picked directly into the shipping carton. When picking is complete, dunnage is added and the carton sealed, eliminating a formal packing operation. This logic works best when picking or packing products with similar size or weight characteristics. In operations with a very diverse product mix, it is much more difficult to get this type of logic to work effectively.

Slotting describes the activities associated with optimising product placement in pick locations in a warehouse. Software packages exist that are designed just for slotting, and many WMS packages will also have slotting functionality. Slotting software will generally use item velocity (times picked), cube usage and minimum pick-face dimensions to determine best location.

Yard management describes the function of managing the contents (inventory) of trailers parked outside the warehouse, or the empty trailers themselves. Yard management is generally associated with cross-docking operations and may include the management of both inbound and outbound trailers.

Some WMSs provide functionality related to labour reporting and capacity planning. Anyone who has worked in manufacturing should be familiar with this type of logic. Basically, planners set up standard labour hours and machine (usually lift trucks) hours per task and set the available labour and machine hours per shift. The WMS will use this information to determine capacity and load. Manufacturing has been using capacity planning for decades with mixed results. The need to factor in efficiency and utilisation to determine rated capacity is an example of the shortcomings of this process.

13.5.5 Implementing WMS

In addition to the standard suggestions of "don't underestimate", "thoroughly test" and "train, train, train", implementation tips that apply to any business software installation, it is important to emphasise that WMSs are very data dependent and restrictive by design. That is, an organisation needs to have all the various data elements in place for the system to function properly. Also, when they are in place, the system must be operated within the set parameters.

Unless the WMS vendor has already created a specific interface with an organisation's accounting or ERP system (such as those provided by an approved business partner), significant amounts of money should expect to be spent on computer programming. While many people hope that integration issues will be magically resolved someday by a standardised interface, this is not practical yet. Ideally, an organisation will want an integrator that has already integrated the selected WMS with the existing business software. Since this is not always possible, an organisation will at least want an integrator that is very familiar with one of the systems.

A lot of other modules are being added to WMS packages. These would include full financials, light manufacturing, transportation management, purchasing and sales order management. Using ERP systems as a point of reference, it is unlikely that this add-on functionality will match the functionality of best-of-breed applications available separately.

13.6 SUMMARY

The correct storage of lubricants is important to maintaining the properties and performance of the products, as well as to improving the cost-effectiveness of the complete supply chain. Lubricant storage is not simply about containers (tank, drums or bottles), but also about warehouses and their operation. Automation and computer control of lubricant storage and warehouse operation is highly likely to maintain product quality, customer responsiveness and supply chain cost-effectiveness.

14 Product Quality Management

14.1 INTRODUCTION

The differences between knowledge and wisdom are very important to thinking about total quality management (TQM). Knowledge is something that can be purchased. It can be obtained by reading books and attending conferences, seminars and training courses. Knowledge remains just knowledge until it is put into action.

Conversely, wisdom is something that is learned by doing. Practice is the best way of learning, and wisdom emerges from practice.

It has been observed that European and U.S. management has tended to stress teaching knowledge in the classroom over wisdom through doing, whereas the Japanese approach for quality management has been to provide both knowledge and wisdom to employees.

This latter approach is particularly effective in solving quality problems on the "shop floor".

14.2 BACKGROUND TO TRUE TOTAL QUALITY

Gemba means "the place where real action occurs". In manufacturing, gemba means "the shop floor". Masaaki Imai, in his book *Gemba Kaizen: A Commonsense, Low-Cost Approach to Management*,[1] illustrated the three major activities to support good gemba management:

- Standardisation.
- Good housekeeping.
- Muda (waste) elimination.

He explained the difference between wisdom and knowledge by citing an example from the housekeeping activities. One of the five steps of good housekeeping in gemba is seiso (cleaning), meaning the involvement of operators in cleaning the machines they work with. As they do so, operators often discover oil leaks or loosening of bolts on the machine. This gives them the opportunity to take corrective actions and eventually develop maintenance standards. This is learning by doing, and the operators gain valuable wisdom about machine maintenance, which is an important step for quality improvement.

There are five golden rules of gemba management:

- When a problem (abnormality) arises, go to gemba first.
- Check with gembutsu (relevant objects).

- Take temporary countermeasures on the spot.
- Find the root cause.
- Standardise to prevent recurrence.

In managing gemba, the most critical part is for managers to go to gemba (the shop floor) and have a good look. Managers who stay away from the shop floor and seldom take the trouble of going there are in contact with the reality of operations only through indirect means, such as reports and conferences. In such cases, managers are making decisions based on fabricated data.

When a manager or supervisor goes to the shop floor where an abnormality occurred, fabricated data are not needed, because what can be observed there is the reality. A manager on the shop floor is right in the midst of reality, and it is likely that the problem can be solved on the spot and in real time simply by following the five golden rules of gemba management.

Another effective approach for problem-solving in gemba has been to collect and analyse data. Generally speaking, when these down-to-earth activities in gemba are carried out, the reject rates should go down to a tenth of their original levels. Unfortunately, many managers in European and North American companies do not take advantage of these effective gemba practices and pursue more academic and sophisticated approaches for quality improvement. These do not work as well as gemba and waste valuable time.

14.3 LEAN (JUST-IN-TIME) MANUFACTURING

Perhaps the most urgent issue currently facing most manufacturing companies is that their present production systems are the biggest hindrance to achieving quality management.

Currently, many manufacturing companies still subscribe to the traditional batch production system. Batch production can be viewed as an antiquated paradigm borrowed from agricultural practices; farm products are sown or born, and then grown, harvested or slaughtered, processed and stored in batches. The more grain in the warehouse or the more sheep in the paddock, the better. Agriculture must take into account the shifting seasons and, for example, it is taken for granted that the lead time of growing and harvesting grain must be long.

When manufacturing began to develop, its processes were modelled on those used in agriculture. Raw materials were bought, processed and stored in batches. Very little consideration was given to establishing a flow of work, and no effort was made to shorten the lead time of production. Keeping a large inventory was taken for granted as a way of doing business. Even now, good inventory means high inventory to some managers.

While the range of different products marketed and sold to customers was small, batch production did not pose many problems. Now that customers demand a wider range of products to be delivered quickly and in different amounts, it has become increasingly difficult to develop the flexibility to meet such demands using batch production. To cope with this, managers have shortened set-up times, added more production lines, introduced more flexible manufacturing and even built new manufacturing plants.

A number of features of the batch production system stand in the way of quality management:

- **Large inventory:** As the name *batch production* suggests, the system is based on producing large batches of inventory at every stage in the production process. As a result, 100% quality control inspection is almost impossible. Even if quality defects are found at a later stage, it is difficult to go back to the previous process that produced the defects, seek out the root cause and take corrective action, since such defective items were made several days earlier. Also, the quality of products or parts deteriorates over time when stored in inventory. (The only exceptions to this, of course, are red wine and whisky.)
- **Long lead time:** The long lead time required by the batch production system makes it difficult to take prompt and flexible action to meet customer requirements for quality and delivery. For example, the batch production system is significantly less flexible when design changes are required.
- **Silo organisation:** Batch production is necessitated because each manufacturing process is separated from each other, each on its own isolated "silo" (a group of people inside a "wall" and separated from other groups in the same company or even in the same building). This necessitates transport between processes, which can cause deterioration or damage. The silos also make it difficult to diagnose quality problems in real time, because different groups do not communicate with each other effectively. When operators do their tasks surrounded by inventory, housekeeping is difficult to maintain, which then can lead to lower morale and less self-discipline.

It is therefore clear that no matter how much effort management may make toward improving quality, batch production impacts those efforts.

The just-in-time (JIT) production system was developed as an antithesis to batch production by Taiichi Ohno at the Toyota Motor Corporation[2] and, together with many other practical tools, such as kanban, poka-yoke (fail-safe device) and jidohka (automation), is supported by the following three pillars of production:

- Takt time versus cycle time: Theoretical time versus actual time for processing one workpiece.
- Pull production versus push production: Producing only as many items as the next process needs versus producing as many as can be produced.
- Establishing production flow: Rearranging equipment layout and processes according to the work sequence.

JIT is really a revolutionary production system and is in every sense just the opposite of batch production. It employs minimum materials, equipment, manpower, utility, space, time and money. It produces products in a shortest lead time, meets the diversified demand of customers and delivers the products when they are required.

However, wide use of the term *JIT* during the 1980s decreased in the 1990s, as the new term *Lean manufacturing* became established. Lean manufacturing is a systematic method for waste minimisation (muda) without sacrificing productivity in a manufacturing system. It also takes into account waste created through unnecessary activities (muri) and waste created through unevenness in workloads (mura). Working from the perspective of the client who consumes a product or service, *value* is any action or process that a customer is willing to pay for.

In Lean, a flow-based approach aims to achieve JIT, by removing variations caused by work scheduling. The effort to achieve JIT exposes many quality problems that are hidden by buffer stocks. Forcing smooth flow of only value-adding steps exposes these problems, enabling them to be dealt with quickly.

Muri is the unnecessary work imposed on employees and machines by managers, as a result of poor organisation. In a lubricant blending plant, muri might be caused when a blending unit operator has to collect a drum of heated additive from a drum heating room before blending can begin. Muri may simply be asking for a greater level of performance from a process than it can handle without taking shortcuts and informally modifying decision criteria. Unreasonable work is almost always a cause of multiple variations in quality.

Linking these concepts is simple in Lean. Muri focuses on the preparation and planning of the process, or what work can be avoided proactively by design. Mura then focuses on how the work design is implemented and the elimination of fluctuation at the scheduling or operations level. Muda is then discovered after the process is in place and can then be dealt with reactively. It is the role of managers to examine the muda in the processes and eliminate the deeper causes by considering the connections to the muri and mura of the system. The muda and mura inconsistencies can then be fed back to the muri, or planning, stage.

Lean manufacturing makes obvious what adds value, by eliminating everything that is not adding value. Although Lean, and before it JIT, started in the automotive industry, it has now been adopted into other industries to promote productivity and efficiency. In global supply chains, information technology (IT) is able to deal with most of Lean practices that synchronise pull systems between suppliers and customers. As a consequence, manufacturers renew and change production strategies and plans just in time.

Quality is ensured by keeping inventories small and through the use of flow production. Small inventories eventually lead to one-piece flow, which is one workpiece moving from process to process. This enables operators to make a 100% inspection of each piece. In flow production, unlike in the isolated islands approach of batch production, processes are arranged in a flow, so any quality reject created in one process can be identified in the next process immediately.

However, Lean and JIT have a number of potential disadvantages. A manufacturer needs to have suppliers that either are close by or can supply materials quickly with limited advance notice. When ordering small quantities of materials, suppliers' minimum order policies may pose a problem. In global supply chains, suppliers that are operating Lean or JIT need to have their suppliers that are able to supply just in time. A global supply chain is as strong as its weakest link. A major disruption in one supplier is likely to have major consequences for everyone else further down the chain. For a lubricant blending plant, the obvious example is the reliable supply of base oils and additives.

Employees are at risk of precarious work when employed by factories that use JIT and flexible production methods. Studies have shown that when employers seek to adjust their work easily in response to supply and demand conditions, they do so by creating more non-standard work arrangements, such as contracting and temporary work.

Another criticism of Lean is that managers may focus on tools and methodologies rather than on philosophy and culture. Effective and focused management is needed in order to avoid failed implementation of Lean methodologies. Another pitfall is that managers decide what solution to use without understanding the true problem and without consulting shop floor staff. As a result, Lean implementations often look good to the manager but fail to improve the situation.

14.4 TOTAL QUALITY MANAGEMENT

Increasingly, lubricant blending companies are finding it important to focus on the distinctive competence they need to have, or to build, to be able to compete. This process is supported by a quality system.

An international standard, ISO 9000, has been established. ISO 9000 is now a requirement in the lubricant industry and provides:

- A common language of quality terms.
- Quality systems elements.
- A model for quality insurance.

Over the last 5 to 10 years, growing attention has been paid to one particular approach: the quality improvement process or TQM to support supply chain management.

The process started in the automotive industry for a number of reasons. The automotive market was mature with an ever-shortening product life cycle. Producers were experiencing ongoing technology changes in the face of a long and cost-intensive development process. Concurrently, Japanese manufacturers began taking significant market share away from their North American and European competitors.

These three factors caused Western manufactures to begin to look at emulating Japanese practices, which included:

- Just-in-time.
- Total quality management.
- Early supplier involvement (ESI).

Following the success of the extensive improvements in the automotive industry, other manufactures adopted of some of these practices.

For example, the quality improvement model involves:

- The role of management.
- Quality improvement process/quality activities.
- Quality system/procedures.
- Relationship with customers.
- Relationship with suppliers.
- Results.

With regard to the relationships between lubricant blenders and customers and between suppliers and lubricant blenders, it is usual to observe that *value chain partnership*, which links buyers to suppliers, requires the strongest, closest and most demanding form of collaboration, compared with joint ventures and consortia. In a value chain partnership, the core business of the partners is to create value for the ultimate user, the customer. The depth of connection is even greater in the area of innovation, traditionally considered the heart of a lubricant business and regarded as confidential. Now, however, lubricant companies are realising that large vertical growth potential results from effective innovation.

Examples of value chain partnership in the lubricant business include:

- Formulations: Additive supplier versus in-house development.
- Production efficiency: External benchmarking, best practice.
- Bottle (package) supplier: Design and capping, e-commerce.

Programmes for TQM have become a standard in order to be competitive in the lubricant marketplace. These kinds of models are continuously raising the bar on quality improvement.

The common standards are:

- Process control and continuous process improvement.
- Customer and supplier relationships.
- Standards for process management.
- Auditing for benchmarking improvements.

TQM is embedded within a quality management system (QMS), which is a collection of business processes focused on consistently meeting customer requirements and enhancing their satisfaction. The QMS must be aligned with an organisation's vision and strategy and is expressed as the goals, policies, processes, documented information and resources needed to implement and maintain it. Early QMSs focused on predictable outcomes of industrial manufacturing, using simple statistics and random sampling. They were then expanded to include employee teams and cooperation, and they then tended to converge with sustainability and transparency initiatives. The ISO 9000 family of standards (see Section 14.5) is probably the most widely implemented QMS worldwide.

Significant amounts of information, both numerical and non-numerical, are required inputs into a QMS. Summaries of these inputs are shown in Tables 14.1 and 14.2.

TABLE 14.1
Information for Quality Management System: Non-Numerical

Information Type	Description
Affinity diagram	Systematic organisation of information to give a clear and objective view of the facts
Benchmarking	Measurement of process against recognised industry leaders
Brainstorming	Generate and evaluate lists of ideas, problems or issues
Cause-and-effect diagram	Systematic analysis of root causes of problems
Flow chart	Description of existing, modified or new processes
Tree diagram	Breakdown of the subject into its basic elements

Source: Pathmaster Marketing Ltd.

TABLE 14.2
Information for Quality Management System: Numerical

Information Type	Description
Control chart	Monitor the performance of a process to determine if its performance reveals normal or out-of-control situations
Histogram	Display the dispersion or spread of data
Pareto diagram	Identify major factors and distinguish the most important causes of quality losses
Scatter diagram	Discover, confirm or display relationships between two sets of data

Source: Pathmaster Marketing Ltd.

Of the numerical information, a histogram is an accurate representation of the distribution of numerical data. It is an estimate of the probability distribution of one continuous variable. A histogram differs from a bar chart, which relates two variables. An example of a histogram is shown in Figure 14.1. The example illustrates that while the weights of most of the blends are relatively accurate, a few are either slightly lower or slightly heavier than planned. This could indicate that either the blend formulation was not followed accurately or the load cells or mass flow meters need recalibrating.

A Pareto diagram, named after Vilfredo Pareto, is a type of chart that contains both bars and a line graph, where individual values are represented in descending order by bars and the cumulative total is represented by the line. The left vertical axis is the frequency of occurrence, but it can alternatively represent cost or another important unit of measure. The right vertical axis is the cumulative percentage of the total number of occurrences, total cost or total of the particular unit of measure. Because the values are in decreasing order, the cumulative function is concave. The purpose of the Pareto diagram is to highlight the most important among a (typically large) set of factors. In quality control, it often represents the most common sources

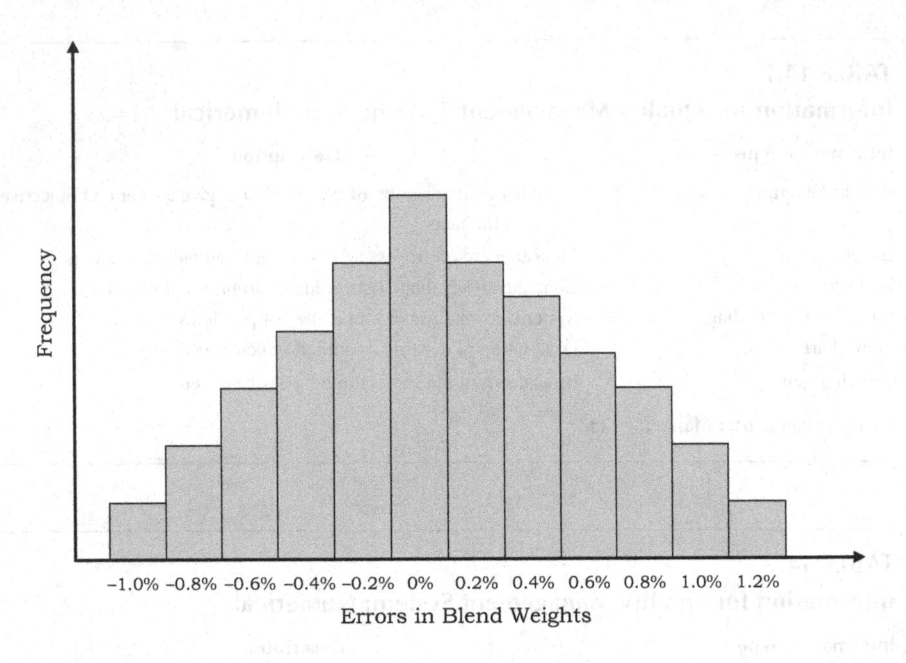

FIGURE 14.1 Example of a histogram.

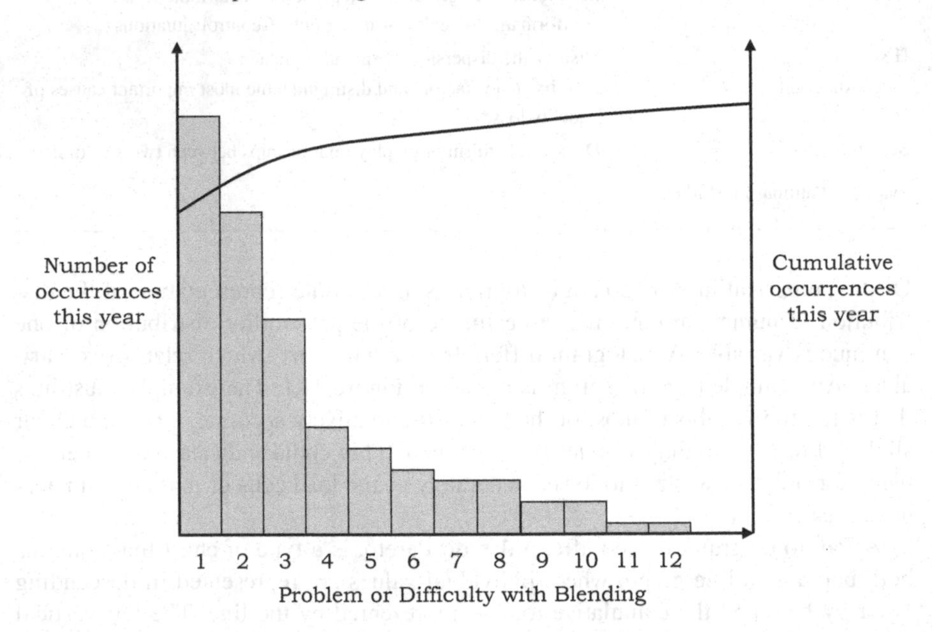

FIGURE 14.2 Example of a Pareto diagram.

of defects, the highest occurring type of defect or the most frequent reasons for customer complaints. An example of a Pareto diagram is shown in Figure 14.2. In the example, problems 1, 2 and 3 with blending appear to be the most important ones to solve or overcome first. The other problems or difficulties could be tackled later.

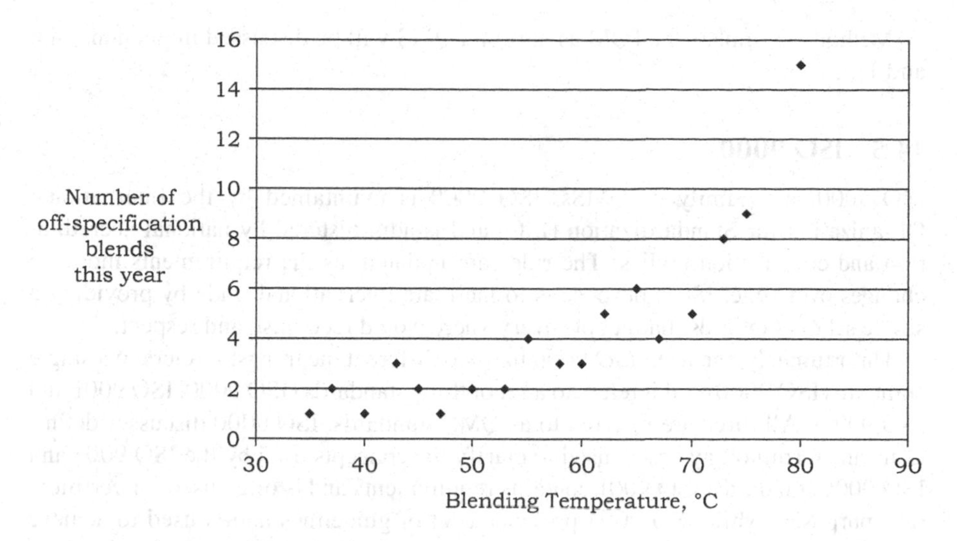

FIGURE 14.3 Example of a scatter diagram.

A scatter diagram is a graph in which the values of two variables are plotted along two axes. The pattern of the resulting points reveals whether any correlation exists between the two variables. If there appears to be a correlation, the points on the graph can be subjected to a regression analysis to determine how strong is the correlation. An example of a scatter diagram which shows very little correlation is shown in Figure 14.3. In the example, although there is not a strong correlation between blending temperature and off-specification blends, there does appear to be a general correlation that the higher the blending temperature, the more likely it is that the blend could be off specification.

The advantages of TQM include:

- It significantly improves the quality of products and services.
- It significantly decreases the waste of resources.
- The productivity of staff increases greatly.
- Improvements in the quality of products and services should lead to increases in market share.
- Employees become more motivated as they achieve their full potential.

At the same time, the disadvantages of TQM can be:

- It demands management and staff time.
- It will only help if the company is heading in the right direction.
- It is not a quick fix; it is an unending process.
- It can lead to too much attention to customers and not enough to employees.
- It can become too bureaucratic and mechanical, leading to an emphasis on continuity rather than continual improvement.
- It can cause disruption at various stages, requiring careful management.

Methods to implement TQM as part of a QMS will be discussed in Sections 14.6 and 14.7.

14.5 ISO 9000

ISO 9000 is a family of QMSs. ISO 9000 is maintained by the International Organization for Standardization (ISO) and is administered by national accreditation and certification bodies. The rules are updated, as the requirements motivate changes over time. ISO's purpose is to facilitate international trade by providing a single set of standards that people everywhere would recognise and respect.

Unfortunately, the term *ISO 9000* has two different meanings: it refers to a single standard (ISO 9000) and it refers to a set of three standards (ISO 9000, ISO 9001 and ISO 9004). All three are referred to as QMS standards. ISO 9000 discusses definitions and terminology and is used to clarify the concepts used by the ISO 9001 and ISO 9004 standards. ISO 9001 contains requirements and is often used for certification purposes, while ISO 9004 presents a set of guidelines and is used to achieve sustained success.

Some of the requirements in ISO 9001:2015 (the latest revision of the standards) include:

- A set of procedures that cover all key processes in the business.
- Monitoring processes to ensure they are effective.
- Keeping adequate records.
- Checking output for defects, with appropriate and corrective action where necessary.
- Regularly reviewing individual processes and the quality system itself for effectiveness.
- Facilitating continual improvement.

ISO 9001:2015 is a major revision to ISO 9001:2008. It is the next regular update, intended to reflect changes in the ways businesses work due to globalisation and more complex supply chains. It has an increased focus on risk-based thinking and greater emphasis on leadership engagement. It helps address organisational risks and opportunities, addresses supply chain management more effectively and is more user-friendly.

According to ISO 9000, the ISO 9001 and 9004 standards are based on eight quality management principles. These principles were chosen because they can be used to improve performance and achieve success:

- **Focus on customers:** Organisations rely on customers. Organisations should understand customer needs, should meet customer requirements and should exceed customer expectations.
- **Provide leadership:** Organisations rely on leaders. Leaders should establish a unity of purpose and set the direction the organisation should take. Leaders should create an environment that encourages people to achieve the organisation's objectives.

- **Involve people in the organisation:** Organisations rely on people, so they should encourage the involvement of people at all levels and should help people to develop and use their abilities.
- **Use a process approach:** Organisations are more efficient and effective when they use a process approach, so they should use a process approach to manage activities and related resources.
- **Take a systems approach:** Organisations are more efficient and effective when they use a systems approach, so they should identify interrelated processes and treat them as a system and should use a systems approach to manage their interrelated processes.
- **Encourage continual improvement:** Organisations are more efficient and effective when they continually try to improve, so they should make a permanent commitment to continually improve their overall performance.
- **Get the facts before deciding:** Organisations perform better when their decisions are based on facts, so they should base decisions on the analysis of factual information and data.
- **Work with suppliers:** Organisations depend on their suppliers to help them create value, so they should maintain a mutually beneficial relationship with their suppliers.

ISO 9001:2015 includes the following main sections:

- Quality management system.
- Management responsibility.
- Resource management.
- Product realisation.
- Measurement analysis and improvement.

A company or organisation that has been independently audited and certified to be in conformance with ISO 9001 may publicly state that it is "ISO 9001 certified" or "ISO 9001 registered".

Certification to an ISO 9001 standard does not guarantee any quality of end products and services. Rather, it certifies that formalised business processes are being applied to the management of quality.

If an organisation already has a functioning QMS, undertaking a gap analysis may be useful. A gap analysis will tell the organisation exactly what it needs to do to meet the ISO 9001 standard. It will help to identify the gaps that exist between the ISO 9001 standard and the organisation's processes. Once the organisation knows where the gaps are, it can take steps to fill those gaps. By following this incremental approach, the organisation will not only comply with the ISO 9001 standard but also improve the overall effectiveness of the organisation's QMS. A gap analysis will also help the organisation to figure out how much time it will take and how much it will cost to bring its QMS into compliance with the ISO 9001 standard.

However, if an organisation does not have a QMS or if it is starting from scratch, an ISO 9001 process-based QMS development plan can be used to develop a QMS. Once the QMS has been fully developed and implemented, the organisation may wish to carry out an internal compliance audit to ensure that it complies with the ISO 9001:2015 requirements. Once the organisation is sure that its QMS is fully compliant, it is ready to ask a registrar (certification body) to audit the effectiveness of its QMS. If the auditors like what they find, they will certify that the QMS has met the ISO's requirements.

While ISO 9001 is specifically designed to be used for certification purposes, an organisation does not have to become certified. ISO does not require formal certification (registration). An organisation can simply establish a compliant **QMS** and then announce to the world that it complies with the ISO 9001 standard. Of course, the compliance claim may have more credibility in the marketplace if an independent registrar has audited the QMS and agrees with the claim.

ISO 9001 is important because of its orientation. While the content itself is useful and important, the content alone does not account for its widespread appeal. Currently, ISO 9001 is supported by national standards bodies from more than 150 countries. This makes it the logical choice for any organisation that does business internationally or that serves customers who demand an international standard of excellence.

ISO 9001 is also important because of its systematic approach. Many people wrongly emphasise motivational and attitudinal factors. The assumption is that quality can only be created if workers are motivated and have the right attitude. This is fine, but it does not go far enough. Unless an organisation institutionalises the right attitude by supporting it with the right policies, procedures, records, technologies, resources and structures, it will never achieve the standards of quality that other organisations seem to be able to achieve. Unless an organisation establishes a quality attitude by creating a QMS, it will never achieve a world-class standard of quality.

Simply put, if an organisation wants to have a quality attitude, it must have a quality system. This is what ISO recognises and is why ISO 9001 is important.

14.6 IMPLEMENTING TQM

TQM is a way of managing that gives everyone in the organisation responsibility for delivering quality to the final customer; quality is described as "fitness for purpose" or as "delighting the customer". TQM views each task in the organisation as fundamentally a process in a customer–supplier relationship with the next process. The aim at each stage is to define and meet the customer's requirements, with the aim of maximising the satisfaction of the final consumer at the lowest possible cost.

TQM has several advantages:

- It vastly improves the quality of the final product or service.
- It greatly decreases the waste of resources.
- Productivity rises sharply as staff use time more effectively.
- As products and services are improved, market share should show a long-term increase, leading to sustained competitive advantage.
- The workforce becomes more motivated as employees realise their full potential.

However, like every other management tool, TQM has some weaknesses:

- It can be extremely demanding of management and staff time.
- It will help only if the organisation is heading in the right direction. It is not a tool for turning an organisation around.
- It is not a quick fix. TQM takes years to implement and, in fact, is an unending process.
- It can lead to too much attention being paid to the needs of final customers and not enough to those of employees.
- It can become overly bureaucratic and mechanical, leading to an emphasis on consistency rather than improvement, or a focus on the means rather than the end.
- It is likely to cause disruption at various stages, requiring careful handling.

An action checklist for implementing TQM is:

- **Decide whether to run pilot programmes:** Although there is a need to map a TQM strategy for the whole organisation, it will usually be introduced in stages. Managers may select for pilot programmes significant areas or functions in which it is considered TQM will yield results within a year at most. Short-term success will be critical in selling TQM to the sceptics; there will always be sceptics.
- **Monitor and evaluate the results of the pilot programmes:** A framework should be defined and a management team appointed to assess and evaluate the results of the pilot programmes. What lessons can be learned and how can these be applied in introducing TQM elsewhere in the organisation?
- **Select tools and techniques to use at each stage in the implementation:** There are four key stages in the implementation of TQM: measurement, process management, problem-solving and corrective action. For each, the tools and techniques appropriate to the scale and environment of the organisation will need to be selected.
- **Select measurement techniques:** Measurement is critical to the success of TQM in quantifying situations and events and providing a benchmark by which to measure progress. The key is to ensure that measurement is a meaningful process leading to corrective action, not an end in itself. The main techniques are measurement and error logging charts, corrective action systems, work process flow charts, run charts and process control charts.
- **Select process management tools:** Many systems and tools can be used in process management. Some, for example, Gantt charts, flow charts and histograms, may already be used in the organisation for other purposes. Those that are right for the specific organisational culture should be selected.
- **Set up mechanisms for problem-solving:** Plan to establish groups throughout the organisation to look at improving quality from different angles. Improvement groups are regular sessions led by supervisors of natural work groups. Key process groups analyse the operation of important processes.

Innovation groups cross departments and are drawn from different levels within the organisation to look at totally new ways of working. The groups should have a range of techniques available to help them, including brainstorming, fishbone diagrams and Pareto analysis.

- **Set up mechanisms for corrective action:** The emphasis in TQM must be on identifying the causes of problems and solving them. At the planning stage, feedback loops with corrective action should be built in.
- **Draw up a communications plan for announcing the TQM programme:** When and how to announce implementation of the programme across the organisation needs to be established. It should be assumed that staff may initially be cynical or sceptical, so strategies for overcoming employees' doubts should be identified. "Converts" from the pilot programmes can be used to explain the benefits. Employees need to be aware of the relationship of TQM to other initiatives within the organisation.
- **Plan to create the right culture for quality:** Successful TQM depends as much on culture change as on process improvements. Senior managers should be aware that TQM will probably need to be accompanied by a general programme of information and education targeted at employees, supervisors and junior managers.
- **Implement the education programme:** The education programme mapped by the strategy can then be introduced. Key groups should be targeted first. These can then be used as agents of change to disseminate learning through the organisation.
- **Empower supervisors:** Team leaders will be pivotal to the success of TQM. They will need to be given the resources, time, support and education to become leaders.
- **Consider how to motivate employees to take ownership:** Employees will need to take ownership of quality and act on their own initiative. To achieve this, an open culture will need to be created. Staff will need to be encouraged that fear of failure, of taking risks and of reprisals should not be part of their thinking. Preparation will be needed to deal with the possible insecurities of managers who discover that most or all of their work is unnecessary or can be done by staff at lower levels.
- **Establish a programme of management change:** Employees will not be able to make the changes needed without profound changes in management style. A new approach based on collaboration, consensus and participation will be needed under TQM. The largest single change for managers will be from telling to listening, from commanding to empowering.
- **Set short- and long-term goals for the implementation programme:** A means for monitoring progress must be established. Short-term goals to demonstrate progress can be combined with more challenging long-term ones to stretch the organisation. A mix of business and cultural indicators can be included.
- **Maintain the impetus:** Cultural changes will take a long time to show results, but without results staff may be frustrated because they do not

perceive much achievement through process improvements. Progress should be reviewed and reported regularly. Successes must be recognised and publicised.

In order to get TQM to work, the following actions are advisable:

- The relationships between TQM and other initiatives within the organisation should be defined.
- The invisible barriers to change should be identified. Awareness of them from the outset and the development of a strategy for breaking through them are vital for success.
- Systems should concentrate on measuring the performance of work processes, rather than the managers, supervisors or employees engaged in them.
- Attention needs to be paid to the soft side of TQM. Changing culture is as important as changing processes.
- Directors and senior managers should make it clear to all staff that TQM is not a quick fix but an ongoing process of continuing improvement. Total quality cannot be fully achieved, as the targets will shift constantly.
- TQM should not be viewed as a precisely defined methodology or a series of neatly tabled sequential actions to be completed one by one.
- TQM should not be introduced at the same time as other major initiatives, if these already make heavy demands on management time.
- The long-term results of TQM must not be obscured by excessive concentration on the short-term means of achieving total quality.

14.7 MAPPING A TQM STRATEGY

An action checklist for mapping a TQM strategy is:

- **Establish a planning team for total quality**: A quality team will be needed to drive through the changes. In a small company, this will be the senior management team; in a larger one, it will comprise senior managers representing the major functions. The team should include known sceptics or mavericks and should ensure that minority views are represented.
- **Assess the need for change:** The competitive position of the organisation needs to be considered. Who the key customers are needs to be established and what they expect of the company needs to be identified. It should not be assumed that the organisation is currently meeting all their requirements. Finding out what customers need is a continuous process, not a one-off exercise. It is instructive to find out how other groups (suppliers, competitors and employees) view the quality of the company's product or service.
- **Define the company's vision:** A vision statement should be created that defines where the company wants to be in terms of serving its customers: this vision should be stretching but attainable. The principles and values

that underpin the vision must be defined. Comparable organisations can be used as models, but the final draft must reflect the company's specific culture and circumstances.

- **Define the standard of service the organisation aims to provide:** The vision needs to be translated into realistic outcomes. What customers, suppliers and employees expect the company to deliver in quality of product or service must be established.
- **Review how closely the company meets its own standards:** There will often be a large gap between customer expectations and reality. The reasons for this across the organisation need to be determined. Key reasons are often external constraints, being let down by suppliers, and internal inefficiencies. It can happen that customers expect too little; their needs, not only their expressed wishes, need to be identified.
- **Audit current levels of waste:** Quality failures must be quantified, by securing from heads of department an audit of current levels of waste. All employees should take part in this audit. Data should be collected as widely as possible, the results costed and the findings presented to the senior management team.
- **Establish the current cost of waste:** How much is currently being spent on rectifying internal failure (for example, reworking of below-quality goods) and external failure (such as handling customer complaints) must be identified. Appraisal costs and the time and money spent on inspection and checking should be included.
- **Decide whether to seek third-party certification:** Whether to include a QMS in the TQM initiative should be decided. If it is, it will lead to third-party certification (ISO 9000 or its equivalent), which may bring benefits with customers and suppliers or even be demanded by them.
- **Define the company's quality strategy:** The results of the waste audit should be used to draw up the quality strategy. This will cover the goals of the strategy, including the revised mission, the systems and tools needed to change processes, the cultural changes needed to create the right environment for quality, details of the resources that can be applied and time frames. Senior management approval of the plan can then be obtained.
- **Draw up a management structure for change:** The culture of the organisation will be critical to the success or failure of TQM. The introduction of team-based working needs to be planned: strong, effective teams are essential.
- **Establish an education and training programme:** Some staff will need training in depth, others less, but everyone should be given a thorough introduction to and familiarisation with what TQM means. Training needs in relation to TQM must be analysed and the additional training required must be costed. The cost can be offset against the expected productivity gains. Plan for general induction and training of all employees in the principles of TQM; coaching of managers, supervisors and team leaders in the soft skills needed to implement TQM; job-specific training in new techniques associated with TQM; and additional training in customer relations.

An external trainer or facilitator is almost always essential, especially in the early stages.

- **Identify opportunities and priorities for improvement:** Priorities for the introduction of TQM must be set. Key processes for early analysis and improvement should be selected. At most, three processes should be chosen initially, with at least one that is likely to demonstrate quick returns in business performance.
- **Establish goals and criteria for success:** Both short- and long-term targets will be needed. Measures of success, in both business and cultural terms, must be established.

This checklist is intended only as an aid to initial thinking and planning. Introducing TQM is a major strategic change that requires considerable research and planning. External advice or assistance is likely to be needed to help implement it.

14.8 SUPPLY CHAIN MANAGEMENT

Supply chain management is the process of planning, implementing and controlling the efficient and cost-effective flow and storage of raw materials, in-process inventory, finished products and related information, from the point(s) of origin to the point of final consumption for the purpose of conforming to a customer's requirements.

A lubricant blending plant is part of the supply chain. Responsibility for managing the lubricant supply chain should lie with the marketing and sales function of the lubricant company. Managing supply chains involve a range of complex activities, including:

- Storage of materials.
- Movement of materials.
- Storage of products.
- Movement of products.
- Flow of information; supplier to customer and customer to supplier.
- Planning.
- Organisation.
- Single logic and implementation.

The process issues associated with supply chain management are the physical flow of materials and products, the organisation and management structures, the information flows and systems and transportation. For a lubricant marketing company, there are numerous options for these process activities:

- **Source of supply:** Products made in one's own blending plant, products made by a contracted third-party blending plant or products purchased or imported from a synthetic lubricant manufacturer.
- **Supply of raw materials:** Base oils from one's own refinery, base oils from a third-party refinery or chemical manufacturer, additives from one or more manufacturers and/or distributors.

- **Mode of supply:** Road, rail, river or sea.
- **Storage:** One's own warehouse, a distributor's warehouse or a joint-venture warehouse.
- **Transportation to customers:** One's own trucks and tankers, a distributor's trucks and/or tankers, contracted rail tank wagons, contracted ISOTainers or flexi-bags, one's own or contracted barges or shared methods of transportation.

Blending plant managers should be intimately involved in the decision-making processes around all these activities.

Within supply chain management, the key issues for logistics and distribution include:

- **Optimum supply mix:** This involves computer modelling and regular optimisation reviews.
- **Primary transportation:** This concerns the movement of products to a local storage depot or warehouse. Competitiveness will relate to the mode of transport.
- **Secondary transportation:** This is dominated by road delivery for most lubricants, whether in trucks or in tankers. In many countries, the environments for delivery to customers are likely to be fragmented. If this is the case, marketing "silos" may cause "turf" issues. In recent years, there have been significant technical developments with the computer-aided planning of delivery routes.

The road transport strategy must be suited to the market the lubricant marketing company is serving, must be capable of implementation for 5 years and must meet market conditions. In many countries, road transportation of chemical products, which includes lubricants, must comply with health, safety and environmental regulations and must be capable of adjusting to changes in those regulations.

If the distribution of lubricants, whether packed or in bulk, is outsourced to a third-party logistics company, issues of customer service, product quality, total costs, safety, control and fragmentation must be dealt with in the contract. The golden rule for outsourcing a key part of the supply chain is "never outsource what you don't understand; outsource the process, not the management".

Over time, a company's supply chain strategy is likely to evolve. Stage 1 typically involves quality/cost teams, longer-term distribution contracts, volume leveraging, supply base consolidation and supplier quality focus. Stage 2 develops to ad hoc supplier alliances, cross-functional sourcing teams, supply base optimisation, international sourcing and cross-location sourcing teams. Stage 3 extends to global sourcing, strategic supplier alliances, supplier TQM development, total cost of ownership, a non-traditional purchase focus, parts and service standardisation and early supplier involvement. Finally, stage 4 involves fully integrated, globally competitive supply

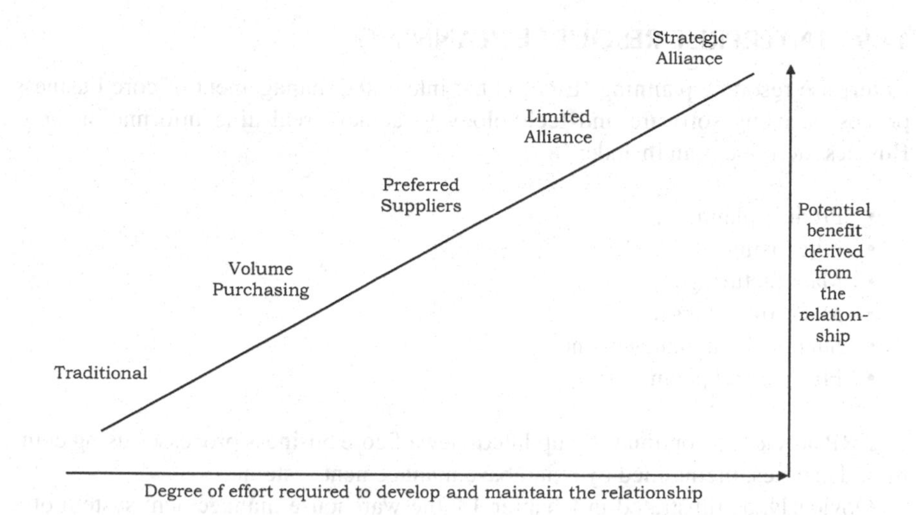

FIGURE 14.4 Illustration of the evolution of business relationships.

chains, cross-company decision-making, management of the total supply chain, full-service suppliers, early sourcing and insourcing and outsourcing to maximise core competencies throughout all supply chains. An illustration of the evolution of supply chain business relationships is shown in Figure 14.4.

As illustrated in Figure 14.4, business relationships can be categorised in five ways:

- **Traditional:** This relationship is typically summarised as "three bids and a buy". It is the historical way of doing business.
- **Volume purchasing:** This involves single sourcing and/or a designated volume of products. Unfortunately, this does not ensure achieving the best price.
- **Preferred suppliers:** A smaller number of suppliers are selected, with specific supply contracts with each. This can lead to shared forecasts and fixed prices.
- **Limited alliance:** This involves a supplier-specific agreement, mutual planning of activities and a business relationship built on trust.
- **Strategic alliance:** This ultimate business relationship also involves a supplier-specific agreement, but one in which the supplier is an integral part of the business. This arrangement gives competitive advantages for both the supplier and customer.

The key benefits of effective supply chain management are the creation and development of supplier–customer partnerships, improved efficiencies for both parties in a partnership and enhanced information flows and control. These lead to process simplification, higher levels of service and significantly lower total costs.

14.9 ENTERPRISE RESOURCE PLANNING

Enterprise resource planning (ERP) is the integrated management of core business processes, using software and technology to achieve real-time information flow. Business activities can include:

- Product planning.
- Purchasing.
- Manufacturing.
- Marketing and sales.
- Supply chain management.
- Finance and payments.

ERP provides a continuously updated view of core business processes using common databases, maintained by a database management system.

Obviously, as illustrated in Chapter 13, the warehouse management system of a lubricant blending plant can be integrated into an ERP system. Functional areas in an ERP system include:

- **Financial accounting:** General ledger, fixed assets payables, receivables, cash management, financial consolidation.
- **Management accounting:** Budgeting, costing, cost management, activity-based costing.
- **Human resources:** Recruitment, training, rostering, payroll, benefits, retirement and pension plans, diversity management, retirement, separation.
- **Manufacturing:** Engineering, bills of materials, work orders, scheduling, capacity, workflow management, quality control, process, projects, manufacturing flow, product life cycle management.
- **Order processing:** Order entry, credit checking, pricing, inventory, sales analysis and reporting, sales commissions.
- **Supply chain management:** Supply chain planning, supplier scheduling, order to cash, purchasing, shipping, claim processing, warehousing.
- **Project management:** Project and resource planning, project costing, billing, performance units activity management.
- **Customer relationship management (CRM):** Marketing and sales, commissions, service, customer contact, call centre support. (CRM is often included in business support services [BSS] rather than in ERP.)

The advantages of an ERP system are numerous. The integration of business process information saves time and cost, allowing management to make decisions faster and with fewer errors. Sales forecasting allows inventory optimisation, transaction history optimises all operations and order tracking aids manufacturing and sales activities. Revenue tracking aids financial control. ERP also eliminates the need to synchronise changes between multiple systems and makes real-time information available to managers anywhere in the company, including the lubricant blending plant. Importantly, ERP protects sensitive data by consolidating multiple security systems into a single structure, which can be better managed as shown in Chapter 8.

However, ERP systems can also have several disadvantages. Customisation of a "standard" system can be problematic, forcing a company to find workarounds to meet unique demands. Reengineering business processes to fit an ERP system can damage competitiveness or divert management focus. ERP costs more than less comprehensive systems. Overcoming resistance to sharing sensitive information between business units or different departments can divert management attention. Integration into an ERP system of truly independent businesses within a larger company can create unnecessary dependencies. Finally, extensive training requirements take resources away from daily operations and can be a huge task.

Fortunately, during the last 15 or so years, ERP systems have been developed and improved. Earlier highly customised ERP suites, in which all parts were highly dependent on each other, should be replaced by a mixture of cloud-based and in-company applications. These can be more loosely coupled and can be exchanged easily as needed. There should be a core ERP that covers the most important business functions, with other functions covered by specialist software that simply extends the core. Different companies can have different core parts and can decide which should be cloud based and which can be in-company behind a firewall. Companies will gain speed and flexibility when reacting to unexpected changes in business processes, due to changing financial or political circumstances.

14.10 SUMMARY

Management of product quality is an integral part of supply chain management. Managing quality does not happen by chance; it needs to be planned and implemented carefully and thoroughly, throughout all levels of an organisation, from the most senior managers to the staff who operate the machines and processes. It even involves the secretaries and cleaners.

Fortunately for all organisations, whether large or small, comprehensive procedures, systems and guidelines now exist to help plan and implement TQM. The ISO 9000 standards are the most extensive international set of standards and advice available. Many companies and consultancies in most countries are now able to assist and audit an organisation's QMS, to certify that it complies with the ISO 9000 requirements.

REFERENCES

1. Imai, Masaaki, *Kaizen Gemba: A Commonsense, Low-Cost Approach to Management*, McGraw Hill Education, New York, 1997.
2. Ohno, Taiichi, *Toyota Production System: Beyond Large-Scale Production* (English translation ed.), Productivity Press, Portland, OR, 1988, pp. 75–76.
3. Imai, Masaaki, *Kaizen Gemba: A Commonsense Approach to a Continuous Improvement Strategy*, McGraw Hill Education, New York, 2012.

Glossary

ABB	Automatic batch blender
ACC	American Chemistry Council
ACEA	Association des Constructeurs Européens d'Automobiles
ADR	Canadian Transport of Dangerous Goods
AFNOR	Association Française de Normalisation
AGMA	American Gear Manufacturers Association
AN	Acid number
ANSI	American National Standards Institute
API	American Petroleum Institute
ASLE	American Society of Lubrication Engineers
ASME	American Society of Mechanical Engineers
ASRS	Automated storage and retrieval system
ASTM	American Society for Testing and Materials
ATIEL	Association Technique de l'Industrie Européene des Lubrifiants
AVG	Automated (Automatic) Guided Vehicle
BN	Base number
BOI	Base oil interchange
BS	Brightstock
CCBL	Cavitation cold blending lubricants
CCS	Cold cranking simulator
CEC	Coordinating European Council
CEN	Comité Européen de Normalisation
CLP	Classification, Labelling and Packaging of Substances and Mixtures (EU Regulations)
CRM	Customer relationship management
DAO	Deasphalted oil
DCS	Distributed control system
DDU	Drum decanting unit
DHU	Drum heating unit
DI	Detergent inhibitor (pack)
DIN	Deutsche Institut für Normun
DOT	U.S. Department of Transport
EC	European Commission
ECHA	European Chemicals Agency
ELGI	European Lubricating Grease Institute
EMA	Engine Manufacturers Association
EOLCS	Engine Oil Licensing and Certification System
ERP	Enterprise resource planning
ESI	Early supplier involvement
EU	European Union
FTIR	Fourier transform infrared (spectroscopy)

GC	Gas chromatography
GHS	Globally Harmonized System of Classification and Labelling of Chemicals
HDPE	High-density polyethylene
HSE	Health, safety and the environment
HVO	High viscosity index
IATA	International Air Transport Association (Dangerous Goods Regulations)
IEC	International Electrotechnical Commission
ILB	In-line blender
ILMA	Independent Lubricant Manufacturers Association
ILSAC	International Lubricant Standardization and Approval Committee
IMDG	International Maritime Dangerous Goods (Code)
IP	Institute of Petroleum (The Energy Institute)
ISA	International Society of Automation
ISO	International Organization for Standardization
JASO	Japanese Automotive Standards Organisation
JIT	Just-in-time
JSA	Japanese Standards Association
KV	Kinematic viscosity
LDPE	Low-density polyethylene
MES	Manufacturing execution system
MRP	Manufacturing resource planning
MRV	Mini-rotary viscometer
MVI	Medium-viscosity index
N	Neutral (base oil)
NLGI	National Lubricating Grease Institute (U.S.)
NPG	Neopentyl glycol
OBS	Optimum batch size
OEM	Original equipment manufacturer
PE	Pentaerythritol
PET	Polyethylene terephthalate
PLC	Programmable logic controller
QMS	Quality management system
REACH	Registration, Evaluation and Authorisation of Chemicals
SAE	Society of Automotive Engineers
SDS	Safety data sheet
SMB	Simultaneous metering blender
SN	Solvent neutral (base oil)
SOP	Standard operating procedure
SUS	Saybolt universal seconds
STLE	Society of Tribologists and Lubrication Engineers
TEOST	Thermal oxidation engine oil simulation test
TMP	Trimethylol propane
TOST	Turbine oil stability test
TQM	Total quality management
UEIL	European Union of Independent Lubricant Manufacturers

UKLA United Kingdom Lubricants Association
VGO Vacuum gas oil
VGRA Viscosity-grade read-across
VHVI Very high viscosity index
VI Viscosity index
VII Viscosity index improver
VLS Verification of Lubricants Specifications (UKLA sponsored)
VM Viscosity modifier
WMS Warehouse management system
ZDDP Zink dialkyldithiophosphate or zinc diaryldithiophosphate

Index

Q

R

S

T

U

V

Printed in the United States
by Baker & Taylor Publisher Services

Printed in the United States
by Baker & Taylor Publisher Services